Against Their Will

NORTH CAROLINA'S STERILIZATION PROGRAM

and the campaign for reparations

Against Their Will

NORTH CAROLINA'S STERILIZATION PROGRAM

and the campaign for reparations

By
KEVIN BEGOS, DANIELLE DEAVER,
JOHN RAILEY AND SCOTT SEXTON

GRAY OAK BOOKS

© 2002, 2003, 2004, 2011, 2012 by the *Winston-Salem Journal*
Preface © 2012 by Paul Lombardo
All rights reserved.
No part of this book may be reproduced in any form without written permission
from the publisher, except by a reviewer who may quote brief passages.

Published in the United States by Gray Oak Books.
For sales and distribution information contact
Gray Oak Books, P.O. Box 404, Apalachicola, FL, 32329
The contents of this book originally appeared in the *Winston-Salem Journal.*
Cover photograph by Ted Richardson.

Library of Congress Control Number: 2011944762
Publisher's Cataloging-in-Publication data
Begos, Kevin Paul.
Against their will : North Carolina's sterilization program /
by Kevin Begos, Danielle Deaver, John Railey and Scott Sexton.
p. cm.
ISBN 978-0-941062-15-2 (hardcover)
ISBN 978-0-941062-16-9 (pbk)

1. Eugenics --North Carolina. 2. Eugenics --United States --History. 3. Involuntary sterilization --North Carolina. 4. Involuntary sterilization --North Carolina --History. 5. Birth control --Government policy --North Carolina --History. 6. African Americans --Medical care --History. 7. Racism --United States --History. I. Deaver, Danielle. II. Railey, John. III. Sexton, Scott. IV. Title.

HV4989 .B44 2012

613.94 --dc22   2011944762

Printed on acid-free paper
Manufactured in the United States of America

Contents

Preface by Paul Lombardo   i

A Note from Executive Editor Carl Crothers   v

Lifting the Curtain on a Shameful Era   1

Still Hiding   15

Tying the Tubes   25

Board Did it's Duty, Quietly   29

Records Unexpectedly Available   33

Fewer People?   34

Forsyth in the Forefront   35

Sign This or Else   47

Benefactor With a Racist Bent   55

Comes a Stranger   61

Wake Forest President Embraced Eugenics Movement   65

Castration: Files Suggest that Punishment was Often the Aim   69

Selling a Solution   71

It Ain't Fair   81

City Kids Put to the Test in '48   87

Newspapers Jumped on Sterilization Bandwagon   93

The Lucky Morons   97

Wicked Silence   103

Detour: In '48 State Singled Out Delinquent Boys 109

Just Carrying Out Orders 113

Church Silent 115

Sterilization was often the Way Out 119

'Bad Girls' Indians Posed a Tricky Race Problem for the State 123

Tempting Choices 127

Stirring up Academia 135

Painless and Permanent 141

Bedfellows in the Cause 145

N.C. Can Still Sterilize Retarded Adults 149

Editorials 151

Epilogue

Easley Apologizes to Sterilization Victims 157

N.C. First to Weigh Eugenics Amends 160

High Hopes: Birth-control clinics opened to fanfare in 1938 163

Some Eugenics Patients Died After Surgery 166

Little Notice and Less Explanation 170

Some Caution Against Ban on Involuntary Sterilization 177

Law that lets judges order sterilizations facing repeal 179

California is latest state to apologize for eugenics 181

Eugenics panel hears of pain 183

Panel accepts change to law   186
House votes 116-1 to end sterilization law   188
Senate votes to repeal sterilization law   190
Class played a role in eugenics sterilizations, researcher says   193
Easley repeals eugenics statute   195
Redress, counsel is aim of project   197
Panel calls for compensating N.C. eugenics victims   199
Offer 'Too Little Too Late'   202
Suggestions abound; Wheels turning slowly   206
Making Amends   208
WFU medical school apologizes again for role   213
Victims of sterilization are still waiting for help from state   217
2010 – State drags feet on promise to sterilization victims   219
2011 – A state with a $25 million fishing pier can help victims   222
2012 – Help for sterilization victims past due   225
2012 – Compensation: Use momentum to map out plan   226
2012 – Tillis, Perdue Could Make History Together   228
2012 – Sterilization Compensation: It fits with state's character   230

Credits   232
Index   234

Preface
by Paul Lombardo

In late 2002 the *Winston-Salem Journal* published an extraordinary series of articles on eugenics, and North Carolina's governor responded by issuing an apology for the abuses of the sterilization program. But even more importantly, the general public was shown for the first time the full scope of how the eugenics program had operated in that state.

It has become commonplace in recent years to describe the history of the eugenics movement in the United States as a story "hiding in plain sight" and that was true in North Carolina.

After all, the 1924 federal law that restricted immigration on eugenic grounds was in force until the mid-1960s, and the so-called "racial integrity" acts prohibiting interracial marriages as threats to the country's gene-pool were not struck down until 1967. Those laws were no secret, but until recently they took up little space in historical accounts and less in the popular press or in the mind of the average citizen.

The laws that permitted more than thirty states to perform sexual sterilization operations on people deemed "feebleminded" or otherwise "defective" were in effect even longer. Two thirds of them remained in place until the 1970s, and the North Carolina law that gave rise to the accounts contained in this book was not repealed until 2003.

Yet while the more than 60,000 surgeries carried out in the name of eugenics have been forgotten by most Americans, they were never far from the memories of the people who endured them. This book tells the stories of people robbed of the potential for parenthood, their lives permanently diminished by eugenic sterilization. It also chronicles a journalistic campaign spanning more than a decade to reawaken dormant memories and shed new light on a dark episode previously buried in our national psyche. It gives a voice, often for the first time, to people previously lost to history.

Scholars have explored the history of eugenics for over a generation, but until recently most academic work has failed to appreciably raise public awareness of past abuses perpetrated in America under eugenic laws. Attention began to shift in 2001, when after significant debate, the Virginia General Assembly voted to adopt a resolution of "profound regret" for state laws passed during the 1920s "eugenics era." Several other states quickly followed with legislation, memorial events and more apologies.

The newspaper series that is reprised in this volume resulted in official recognition by the North Carolina legislature of the state's involvement in the eugenic sterilization movement and apologies by Wake Forest University and the *Winston-Salem Journal* itself, for the paper's strong support of eugenics in decades past.

North Carolina became the first—and to date, the only state—to take serious steps toward compensating the living survivors of sterilization laws, and that debate may be reaching a turning point in 2012.

But North Carolina was hardly unique. Those laws were enacted by legislatures in most of the states in this country, and were on the books in some cases for more than seventy years. A common question raised by people hearing this for the first time is: why were those laws ever passed? It requires no special training to understand that people who live on the lowest rungs of the social ladder are often invisible, and laws targeting them gained legislative approval with little opposition. The people marked for sterilization were people living with disabilities, poor people, often people of color, in short, any of the least favored members of our society.

Those were always the most likely victims of eugenics, and as often as not, the argument used to justify the "surgical solution" of sterilization was a simple one. Preventing childbirth among society's least wanted citizens, eugenicists said, would prevent more people like them being born to fill up the prisons, and hospitals and welfare rolls.

From the beginning of the eugenics movement in America, fears of a decline in population quality went hand in hand with fears of rising taxes. Much later, and well after World War II, when eugenics had come to be identified with Nazi genocide, many states abandoned their sterilization programs. But North Carolina and a few others accelerated the pace of surgery at the same time precisely to focus on illegitimate births. Women on welfare became the usual suspects for sterilization.

The North Carolina program ended in 1974, and the law was officially rescinded twenty-nine years later. When laws are repealed, they leave few traces in the statute books. Yet the scars they have etched into the bodies of the sterilized do not simply disappear. This book is an attempt to insure that the memories of American eugenics, and the memories of its victims, do not merely fade away; it will make the testimony of people who endured eugenic sterilization part of a permanent record. As we remember those people, we can only hope that the misguided efforts to use eugenic policies to scrub the national genome clean will also remain part of the national consciousness.

The book's fundamental message is contained in its title: *Against Their Will*. The operations done on North Carolinians violated a principle established in America before the Constitution was written: no one should be forced to endure surgery without consent. This book reasserts that principle as a reminder for the future.

Paul A. Lombardo

Georgia State University College of Law

author of

Three Generations, No Imbeciles: Eugenics, the Supreme Court & Buck v. Bell

December, 2011

A Note From Carl Crothers
EXECUTIVE EDITOR

It was for their own good, the government said again and again. Nine out of 10 sterilization petitions reviewed by the Eugenics Board of North Carolina during its 40 years of existence were approved, in all more than 7,600 cases. Many were institutionalized mental patients, but most were not. By the time the program was halted in 1974, several thousand citizens had been sterilized under the pretext that it was good for them and good for society.

They were unfit to reproduce, according to the government, for being poor and illiterate, hypersexual or homosexual, promiscuous, even lazy. These and other undesirable characteristics led them to be classified as feebleminded. What followed was swift and irreversible.

That the eugenics movement in North Carolina, as in more than 30 other states, was ultimately ruled by history as misguided does not begin to tell the full story.

Today, the *Winston-Salem Journal* begins a five-part series, *Against Their Will*, that examines in stunning detail North Carolina's eugenics program, whose goal was the systematic elimination of undesirable genetic strains in the population.

Journal reporters Kevin Begos, John Railey and Danielle Deaver obtained thousands of documents, reports and minutes pertaining to the eugenics board, many containing chilling exchanges between board members and citizens called before it.

The records, shielded from public view by the state, show that while other states were shutting down their programs after the revelations of Hitler's crimes in World War II, North Carolina expanded its program to include the general population. Almost two-thirds of the 7,600 North Carolina citizens sterilized had never been institutionalized, and 2,000 were under 19.

Winston-Salem, through the efforts of some of the city's most prominent citizens and a researcher at the Bowman Gray School of Medicine, led an effort to expand the eugenics program. Indeed, this newspaper endorsed the program with positive stories and editorials.

Perhaps most disturbing, the program took a racist turn in the 1950s and 60s, targeting young black females with out-of-wedlock children for sterilization over any other group. Those involved deny a racist motivation, but the evidence is compelling that racial attitudes of the period influenced decisions. We expect criticism for bringing these details to light after so many years. But this is an important story, and one we felt obligated to write.

We Americans tend to ignore our past, wrote filmmaker Ken Burns. Perhaps we fear having one, and so burn it behind us like so much rocket fuel, forever looking forward, forever condemned to repeat that which we did not care to examine.

Lifting the Curtain on a Shameful Era

Thousands were sentenced to sterilization during rubber-stamp hearings in Raleigh

By Kevin Begos
JOURNAL REPORTER

They were wives and daughters. Sisters. Unwed mothers. Children. Even a 10-year-old boy. Some were blind or mentally retarded. Toward the end they were mostly black and poor. North Carolina sterilized them all, more than 7,600 people.

For more than 40 years North Carolina ran one of the nation's largest and most aggressive sterilization programs. It expanded after World War II, even as most other states pulled back in light of the horrors of Hitler's Germany.

Contrary to common belief, many of the thousands marked for sterilization were ordinary citizens, many of them young women guilty of nothing worse than engaging in premarital sex.

I don't want it. I don't approve of it, sir. I don't want a sterilize operation.... Let me go home, see if I get along all right. Have mercy on me and let me do that.
— A woman pleading with the eugenics board, 1945.

The sterilization program ended in 1974, but its legacy will not go away. Many of its victims are still alive and they bear witness to a bureaucracy that trampled on the rights of the poor and the powerless.

In response to a *Winston-Salem Journal* investigation of the state sterilization program, the Wake Forest University School of Medicine is looking into its own role in the eugenics movement.

The state program was run by the Eugenics Board of North Carolina, a panel of five bureaucrats who usually decided cases in a few minutes. It was inspired by the eugenics movement, which made exaggerated claims that mental illness, genetic defects and social ills could be eliminated by sterilization. The system granted excessive power to welfare workers, browbeat women into being sterilized and had ineffective safeguards.

"They don't want to hear how I feel, or what's going on in my mind. You're pregnant – you need to get sterilization," said Nial Cox Ramirez, recalling her sterilization in 1965 after having one out-of-wedlock child.

"And they had the nerve to tell me, "That's what's best for you,'" she said recently.

North Carolina sealed most records of the eugenics board and until recently few details were known about how the board operated, or the nature of cases it handled.

The *Winston-Salem Journal* obtained and examined thousands of these documents. It found that:

More than 2,000 people ages 18 and younger were sterilized in many questionable cases, including a 10-year-old who was castrated. Children were sterilized over the objections of their parents, and the consent process was often a sham.

The program had been racially balanced in the early years, but by the late 1960s more than 60 percent of those sterilized were black, and 99 percent were female.

Doctors performed sterilizations without authorization and the eugenics board backdated approval. Forsyth County engaged in an illegal sterilization campaign beyond the state program.

Major eugenics research at Wake Forest University was paid for by a patron whose long history of ties to science had a racial agenda that included a visit to a 1935 Nazi eugenics conference and extensive efforts to overturn key civil-rights legislation.

North Carolina's eugenics law, passed in 1929 and rewritten in 1933, allowed sterilizations for three reasons – epilepsy, sickness and feeble-mindedness. But the board almost routinely violated the spirit, if not the letter, of the law by passing judgment on many other things, from promiscuity to homosexuality.

Though more than 30 states had eugenic sterilization programs, North Carolina's record of dramatically expanding the program after 1945 and targeting blacks in the general population was different from most.

"That's quite astounding," said Steve Selden, a professor at the University of Maryland and author of *Inheriting Shame: the Story of Eugenics & Racism in America*. "It's simply a story that has not been told."

Paul Lombardo, the head of the University of Virginia's Center for Biomedical Ethics, said that North Carolina was "a unique example." Others have wondered why that was the case.

"Why did it happen in North Carolina and not elsewhere?" asked Daniel Kevles, a professor of history at Yale University and the author of *In the Name of Eugenics.*

One reason was a group of Winston-Salem's elite who formed the Human Betterment League in 1947. Hosiery king James G. Hanes and Alice Shelton Gray, a trained nurse and another member of the local elite, joined forces with Dr. Clarence Gamble of Boston, the heir to the Procter & Gamble fortune. The group launched a massive publicity campaign in North Carolina to promote sterilization programs. Newspapers – including the Journal – bought into it, asked few hard questions, and paved the way for the eugenics board to expand its activities.

"I regret that the Journal, in its past, played a role in legitimizing these barbaric activities," said publisher Jon Witherspoon. "On behalf of the Journal, I apologize for the paper's part in depriving these individuals of their basic human rights."

The rich helped bankroll the program's expansion, and part of the motivation was financial. Awash with inherited money, Hanes and Gamble were concerned about how much welfare mothers and the mentally ill were costing taxpayers.

The eugenics board's files provide another answer to what happened in North Carolina, said Johanna Schoen, an assistant professor of women's history at the University of Iowa who gave the Journal access to a set of 7,000 records that she was allowed to copy more than 10 years ago. Since that time, the N.C. State Archives has declined other requests, and the records are officially closed to the public. The Journal has honored the medical confidentiality of the records.

"This view that we often have of sterilization – and particularly eugenic sterilization – of just being this evil thing that the state does got extremely complicated once I was confronted with these individual stories," said Schoen.

"There are stories of the state doing incredible evil, and then there are stories of women who really want the sterilization, and then there are stories of women and men who are so mentally ill that they really are totally unable to take care of children ... there are no other solutions," said Schoen, who has a book on the history of birth control and sterilization forthcoming from the University of North Carolina Press.

The program may have been complex, she said, but one thing is clear.

"I think the problem is that there are cases where sterilization was the solution – but sterilization authorized by the eugenics board is never the solution," she said. "The very premise that the state had the right to do this was flawed."

For decades there was little public debate over the program, and it went on operating so quietly that few public officials looked into what the board was doing.

Chris Coley of Raleigh was a staff lawyer with the N.C. attorney general's office in the early 1960s. He attended eugenics board meetings with few reservations. "But later on in life, reflecting back on what I was doing, I was a little bit shocked that there was such a procedure."

Those who were swept up in the program suffered through the state's flawed premise.

You are absolutely tearing down the laws of God when you do this. God said when he drove Adam and Eve out of the Garden of Eden, — did He tell her to go out and be sterilized? God said go out and multiply. If sterilization isn't against that, what is?
— The grandfather of a 17-year-old girl, eugenics board hearing, 1938

The shadows of science

Good many of these farmers raise hogs and practice that on the female hogs. Spraying (sic) I think they call it. Same type of operation but they don't remove anything like they do in hog.
— Dr. J.C. Knox, trying to reassure a family protesting
the sterilization order for a 13-year-old girl, 1938

The eugenics movement claimed that human traits such as intelligence, sexuality and criminality were determined almost entirely by genes, or "good blood." "It was misguided to assume that all behavior is directly the result of a gene," Kevles said. "There are a large number of these things that are multigenetic if they are genetic at all, and it was evident that was the case by the 1920s."

The idea flourished, said Lombardo of the University of Virginia, because it suggested that science could provide a simple solution to complex medical and social problems that have been a part of human existence since the dawn of civilization.

Eliminate the "bad" genes from the population, and future generations would flourish, the eugenics movement claimed – a rallying cry that helped inspire Hitler's idea of a master race. "It's hopeful, which is why it was so popular. That was the seductive part of it," Lombardo said.

Earlier this year, Virginia became the first state to issue a statement of regret for its sterilization program. The governor of Oregon apologized for a similar program last week. California led the nation with more than

21,000 sterilizations; Virginia was second with about 8,000, and North Carolina third. Many other states that had sterilization programs have lost the records or, in the case of Oregon, destroyed some of them.

The eugenics movement had so many scientific, medical and legal flaws, Lombardo said, that the idea of denying all access to the material is wrong.

"It does seem clear to me that when the procedures weren't appropriately followed, and people either couldn't understand or didn't know they were being sterilized, states do have an obligation to admit the truth," he said.

There were warnings from scientists and religious leaders when the program was in its infancy, but North Carolina ignored them.

"Those who act in this way are at fault in losing sight of the fact that the family is more sacred than the state," Pope Pius XI wrote in a 1931 statement protesting the eugenics movement. "Public Magistrates have no direct power over the bodies of their subjects."

Inside the room

Thunderstorms rolled across the state and the temperature hit 90 degrees in Raleigh on July 25, 1945. The papers were full of news about the final drive to destroy the Japanese navy in the Pacific, and in previous months the horrible truth of Nazi death camps had shocked even a war-weary world.

There would have been no public mention of the smaller battle that was about to unfold in a small room in a state building. The eugenics board still did not have either a permanent office or a full-time secretary.

The board was considering the case of Sally, a 30-year-old white woman in a state institution.

"As I understand this case, it was considered as a consent case in the beginning," said Clifford Beckwith, representing the state attorney general, as the hearing began. "I understand now that there is a misunderstanding about that. You were not quite sure what you had signed. Is that so?"

"Yes, sir," replied Sally's parents.

Beckwith told the group, which included a sister and a cousin, that everybody would have an opportunity to speak and that the "misunderstanding" would be cleared up. "We are all working for the same thing – to do the best that can be done for the young lady and for all of you."

"The misunderstanding was that they did not give it to me right," Sally's father said. "I can't read nor write or understand writing and went ahead and signed. She's needed at home. No help with the victory garden so that it'll soon be gone. She ain't never been crazy."

"We are working to do the very best for her," Beckwith replied.

"As I understand the case, the patient has recovered from her nervous condition at the present time," Dr. F.L. Welpley said. "She might have recurrent attacks, and if she married and had children they would have a tendency to have some kind of nervous and mental trouble through heredity. The idea would be to operate in such a way as to merely prevent her from having children. It would in no way affect her social life."

"Sally has never been a filthy girl," Sally's mother said. "She has just been overworked more than anything else."

Dr. R.T. Stimpson asked if Sally would be able to care for children, and her mother replied, "She is the best child I have got, and let me tell you, a mother that has raised a large family don't want their children sterile — because I know she don't need it."

"We are looking after the welfare of the patient and the public, too," Welpley added. "If Sally had children, two or three might have to go to institutions."

"I don't see why she needs sterilizing," her cousin said. "She stays at home and works all the time. Sent her over here (the hospital) because she worked too hard."

"I never knew hard work made people nervous," Stimpson said.

"You just never done any," the cousin shot back. "Try it and see."

The hearing dragged on for perhaps 45 minutes and, like most other hearing cases, it seems to have been only a formality.

"Is there anything further to say?" Beckwith asked.

"If you would stop this and let me go home," Sally replied. "Have mercy on me and let me do that."

"What is decided will be our best judgment," Beckwith said.

After the hearing the eugenics board voted 4-1 that Sally should be sterilized before release from the state institution.

Personal stories

The story of the eugenic sterilization movement in America has usually been told with sketchy, impersonal statistics, but the Journal's review of the North Carolina records presents the first overview of a eugenics bureaucracy from beginning to end and an inside look at the human tragedy that was taking place behind closed doors.

Few of the people who were sterilized ever appeared before the eugenics board. Officials consistently led the public to believe that the program did not force or pressure people to have sterilizations. But like Sally's family, most who objected did so in vain.

It would ruin her for life. We don't know whether she is ever going to have (children).
— A mother protesting the sterilization order for her 17-year-old daughter, 1938

We just can't see, Mrs. —, that you have given us any real reason why it shouldn't be done.
— Reply of board member Paul F. Mickey, representing the state attorney general

White and black, male and female, people were deemed to "need" sterilization for their own good or for the good of society. The individuals were vastly different, but the votes usually weren't — the eugenics board approved more than 90 percent of the petitions it reviewed.

Lifting the Curtain on a Shameful Era 9

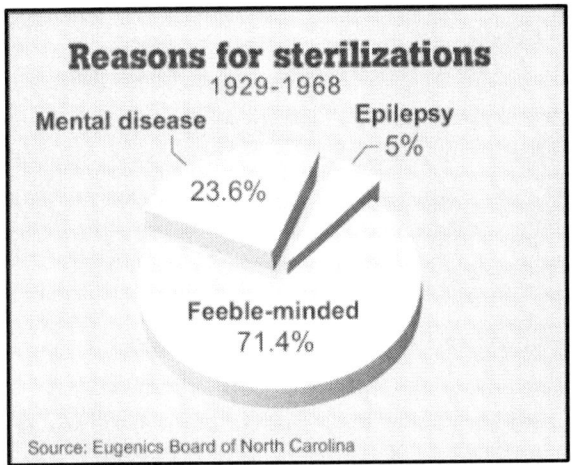

From beginning to end, the records are filled with casual comments, not serious medical discussions.

"Pauper. Needs close supervision. Hypersexuality," reads part of a 1939 petition that reduced a woman's life to 24 words. Still, that was enough for the eugenics board to conclude that she was feebleminded.

Other excerpts from the minutes read:

She wears men's clothing all time (sic), (1947)

Her mother says that she did not go to school as regularly as she could because she had sleepy spells and slight attacks of epilepsy...., (1954)

She seems lazy and unconcerned, (1960)

... while in school attempted to write love letters to boys she imagined were interested in her, (1962)

The entries got longer by the 1960s, but the extra words were hardly a sign of an improved program. In one case a father suspected of incest asked the eugenics board to sterilize his daughter, as if that were a solution.

This fourteen year old girl lives in a very poor home environment. Both parents appear to be limited, and the father admits to incestuous feelings for (his daughter) to his wife. Mrs.___ has been reported to the agency for sexual promiscuity by her own daughter but does make some efforts to give supervision. After the father admitted his feelings for ___ the mother had ___ carefully examined by a physician who reported that she had had intercourse.... The parents wish sterilization for ___ as they are afraid she will become pregnant. Consent: signed by client's father, ___. — Eugenics board records, 1962

The eugenics board approved the sterilization.
Why?
Schoen said she thinks that amidst hundreds of sad stories in the records are part of the explanation for why the North Carolina program continued long after the core scientific and social principles had been rejected in other states. At almost every meeting, she said, there were some cases that appeared to be justified or that presented situations that were much harder to manage before birth control was widely available.

Single illiterate girl, 20 years of age, who is the mother of two children.... At the age of 15 she cut her father in the head with an axe, and he died from injuries received. She shot her brother in the arm which necessitated the amputation of his arm above the elbow. At still another time she poured kerosene on her sister's hand and ignited it...." — Eugenics board minutes, 1949

This 32 year old woman is in her tenth out of wedlock pregnancy, and has seven living children. She loves her children and gives fairly adequate physical care but depends on her father to discipline the children. ... She is quite anxious for the operation as she thinks this is the only way she can stop having children. The parents agree to this. — Eugenics board minutes, 1962

___ is a patient at ___ Hospital for the fourth time. She has been unable to assume the responsibility for her family since the children were placed in foster care in April 1960. The husband is in and out of prison.... When released from the hospital, ___ has returned to her old pattern of prostitution. Since there are

so few strengths in the family, sterilization will prevent additional children being born for whom there is no care. — Eugenics board minutes, 1963

Schoen thinks that the cases of serious mental illness or of people who truly wanted the operation allowed board members to justify their work and overlook the many cases that were far more complicated. After a close examination of more than 7,000 records, Schoen said, she found just 446 cases in which the patient clearly desired the operation.

In thousands of the cases, the records are so sparse or unreliable that it is impossible to tell whether the patient was sterilized willingly or not. While almost all of the petitions showed the "consent" of a relative, patient or guardian, Schoen said, "You can't talk about this consent being freely given." Patients in state institutions were told that they had to agree to sterilization as a condition of release, and in many cases people on welfare were threatened with loss of benefits, she said.

At one meeting the eugenics board took the official position that approval for operations could not be backdated, but in several other cases it did just that. Reports that many doctors were doing large numbers of sterilizations without approval did not result in investigations or reprimand, even though there was no voluntary sterilization law in North Carolina for individuals until 1963.

In the case of _____, Harnett County, where there had been a change of surgeon not authorized by the Board and the operation of sterilization performed on a date prior to the action of the Board, the Board authorized a nunc pro tunc order (now for then) be issued to include the date the operation was actually performed.... — Eugenics board minutes, 1959

The medically and ethically flawed case summaries meant that even well intentioned eugenics board members could make wrong decisions, Schoen said.

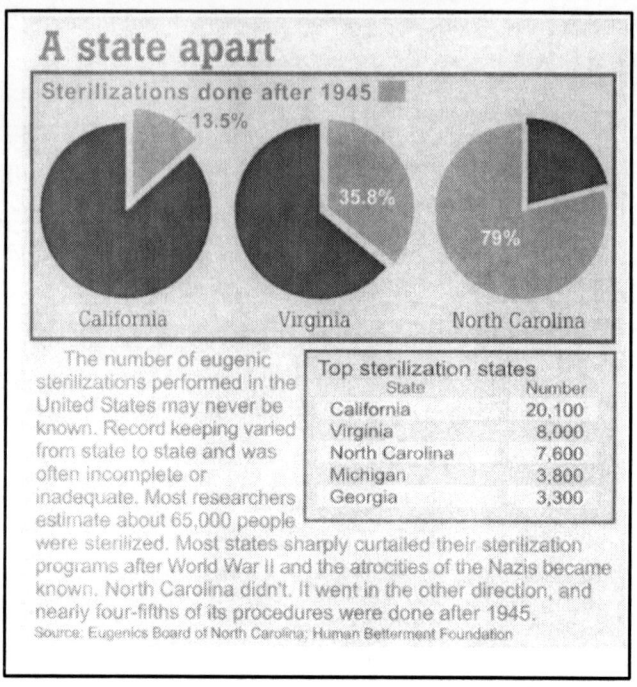

A state apart

Sterilizations done after 1945

California — 13.5%
Virginia — 35.8%
North Carolina — 79%

The number of eugenic sterilizations performed in the United States may never be known. Record keeping varied from state to state and was often incomplete or inadequate. Most researchers estimate about 65,000 people were sterilized. Most states sharply curtailed their sterilization programs after World War II and the atrocities of the Nazis became known. North Carolina didn't. It went in the other direction, and nearly four-fifths of its procedures were done after 1945.

Top sterilization states

State	Number
California	20,100
Virginia	8,000
North Carolina	7,600
Michigan	3,800
Georgia	3,300

Source: Eugenics Board of North Carolina; Human Betterment Foundation

The lack of reliable data came in part from the fact that North Carolina was the only state where social workers had the power to recommend sterilization. With little or no medical training, such workers were poorly equipped to judge complex situations. Some counties did large numbers of sterilizations, while others did almost none. Whether people were sterilized often revolved around the attitude of an individual social worker.

"I never participated, that I recall, in a training course about sterilization," said Ed Chapin, a former director of the Mecklenburg County Department of Public Welfare who handled sterilization cases in the 1960s. "It was just something you picked up."

Some social workers got carried away, he said. One of his co-workers started sterilization petitions for almost "his entire caseload," which was 60 women, Chapin said.

"It was clearly a great option for some people," he said. "There were some kids ... who were genuinely retarded, and limited. And in those cases it was really a blessing for the family. And there wasn't a lot of resistance to that type of situation.

"But then on the other side of the coin, I think there were individuals who kind of saw this as "Enough's enough, she's already had one or two children, and let's put a stop to it.' Those people were kind of, I think, taken advantage of," Chapin said.

But for most of the history of the program, there was no public acknowledgment from officials that the issues were complicated and that the state had failed to carry out proper safeguards against abuse. Newspapers and medical journals consistently failed to take hard looks at what was happening. Instead, they chose to repeat official assurances that problems didn't exist.

Medical and legal experts say that a debate over the eugenics movement isn't just one for the history books.

"The ethical issues that were raised by eugenics are likely to be the very same ethical issues that are being raised with genetic research, now and in the future," said Selden, the University of Maryland professor. "They didn't have the technology to achieve their goals," he said of the eugenics movement. "We do."

'Still Hiding'
Woman sterilized at 14 carries a load of shame

By John Railey and Kevin Begos
JOURNAL REPORTERS

ATLANTA– Elaine Riddick Jessie can't forgive the state of North Carolina for what it did to her in an Edenton hospital in 1968. She tenses as she talks about being sterilized soon after delivering her first and only child when she was 14.

"I was just a baby," said Jessie, 48. "I (was) just a child. They did not, could not have gotten my permission because I wasn't old enough."

Her story isn't one of an isolated mistake. The Eugenics Board of North Carolina authorized sterilization for more than 2,000 girls and boys ages 18 and under between 1933 and 1974; some of them were just 10 years old.

Jessie, who lives in Atlanta now, got no explanation before or after the operation, much less follow-up counseling. She was left with a physical and emotional burden that she is still trying to come to grips with almost 35 years later.

"Why didn't they just sew me up, just sew me up, period? I felt like I didn't have a sex .... I wasn't a male and I wasn't a female. Just asexual. I didn't have a sex, because if I was a woman I could have children," she said. "I hide. I hid. I think I'm sort of still hiding, but there's nothing I can do. It made me dislike myself. And I don't ever think I can like myself. It is the most degrading thing, the most humiliating thing a person can do to a person is to take away a God-given right."

The United States would land men on the moon the year after Jessie was sterilized, but North Carolina was still clinging to a eugenics program that used long-discredited science from the 1920s as its foundation.

The decision in Jessie's case - just as in thousands of others like hers – was made by five strangers in Raleigh who reviewed a few paragraphs that reduced her complex life and the profound medical, legal and ethical issues at stake into a synopsis that seemed to always result in the same answer.

About 90 percent of sterilization petitions presented to the eugenics board were approved, and most cases were decided in less than 15 minutes.

The system was flawed from the beginning, but by the time it caught up to Jessie it had been overwhelmed by sexual and racial bias.

The biannual eugenics board report for 1966 to 1968 shows that 99 percent of the operations were performed on women; 64 percent of those were on black women. From 1929 to 1940, the ratio had been almost the opposite - an overall racial split of 79 percent white and 21 percent black.

The state categorized Jessie as "feebleminded" as part of its justification, but she wasn't an inmate at a state institution. She was just a poor young black girl, like the majority of people who were sterilized in the 1960s. Officials also justified their action by terming Jessie promiscuous.

Jessie said she was not feebleminded or promiscuous. A man in his 20s coerced her into a sexual relationship, she said, and impregnated her when she was a minor - statutory rape by law.

Poor and under pressure

At 5 feet, 2 inches tall and 115 pounds, Jessie is about the same size as she was when she was sterilized. She grew up in Winfall, a tiny town in Perquimans County near Edenton. It's a place where flat fields of cotton and peanuts fall off to the Albemarle Sound. It's a beautiful but hard place, one where many residents still live in crushing poverty.

Jessie remembers her large family sitting on their run-down porch, eating a "hoecake" and beans out of one pot. "We were like ... God, what were we like? We didn't know anything about plates and spoons," she said.

The Perquimans County Department of Public Welfare took custody of Jessie and her seven siblings. They sent five of the children to an orphanage an hour's drive away in Oxford. Jessie and another sister were sent to live with their grandmother, just down the street from their parents' home. That house, too, was dilapidated and crowded.

Shortly after moving in with her grandmother, Jessie said, an older man impregnated her. She was 13.

"All I remember is that it hurt," she said.

Jessie's grandmother, Maggie "Miss Peaches" Woodard, was on welfare, and her social worker learned during a routine visit that Jessie was pregnant. The white social worker with the county welfare department, the late Marion Payne, pressed Woodard to consent to have Jessie sterilized. Finally, Woodard, who is illiterate, signed her "X" on a consent form.

"I didn't know what I was signing," Woodard, 86, said recently. But she said Payne told her that if she didn't sign, Jessie would have to go live in an orphanage.

Jessie's family wasn't the only one to face such pressure. Records kept by the eugenics board show a long history of county officials who treated the issue of consent as more of a nuisance than a moral imperative.

In a 1950 book on the sterilization program written by researcher Moya Woodside, state officials complained about "the necessity of obtaining consent from relatives" who might be of "low mentality" themselves and the "further difficulty" of obtaining "informal consent of the client."

That attitude was still present in the 1960s.

(The) chief problem is securing consents and getting patients to have the operation once the order is issued by the Eugenics Board. One (social) worker said she wishes there was a law which would permit the use of force in getting the operation performed once the order was received.

Jessie's father also signed a consent form. He was a shell-shocked veteran of World War II and drank heavily, as did Jessie's mother. "They were drunk all the time, fighting all the time or either in prison," Jessie said.

She still can't understand why her father had any say-so in her life. "If you're not qualified to take care of your children, how are you qualified to sign papers? They knew he had... mental problems," she said.

In Jessie's case and in many others throughout the history of the program, blatant warnings of ethical problems in the consent process were ignored.

Single girl, 17 years of age.... Her mother is deceased. Her father is an excessive drinker. The father was suspected of incest.... sterilization was ordered on condition that the written consent of the father be secured.
– From a 1948 sterilization petition

She comes from a home where the father is known to indulge in alcohol and has molested ____ and attempted to molest a younger daughter in the home. Consent:____,father. – From a 1968 sterilization petition

A key part of the process in Jessie's case was the evaluation by psychologist Helton McAndrew, who wrote that Jessie was in "the slow section" of her class. She noted Jessie's score on an IQ test - 75.

"(Jessie's) chief problem is her poor home .... We expect this girl to perform more adequately in an improved environment, but it may be desirable to think about vocational training in the future," McAndrew wrote.

McAndrew stopped short of classifying Jessie as "retarded." But Dr. David Wright of Edenton, who hadn't evaluated Jessie, did use that classification in an affidavit included with the petition. Wright, who still lives in Edenton, declined to comment for this story.

As her pregnancy advanced, Jessie dropped out of the eighth grade. Meanwhile, the petition to have her sterilized had made its way to Raleigh, the home of the eugenics board.

A life's summary

When such petitions arrived from across the state, the executive secretary of the board would read them and boil them down to a brief summary for the board to consider.

Jessie's read, in part:

8. Delores Elaine Riddick - (N) - Perquimans County
Social information: Age 13. Single. Pregnant. Psychological
April 5, 1967. MA 9-6: IQ 75
This thirteen year old girl expects her first child in March 1968....She has never done any work and gets along so poorly with others that her school experience was poor. Because of Elaine's inability to control herself, and her promiscuity - there are community reports of her "running around" and out late at night unchaperoned, the physician has advised sterilization....This will at least prevent additional children from being born to this child who cannot care for herself, and can never function in any way as a parent.
Diagnosis: Feebleminded

The summary omitted the section of McAndrew's report dealing with the impact of Jessie's environment on her IQ.

And there was another glaring problem, too. Flawed as the IQ tests of that time were, the eugenics board ignored its own longtime standard that only those who scored below 70 should be sterilized.

Jessie's IQ score was the same as another girl who was spared, yet Jessie's case continued forward.

On Jan. 23, 1968, in a conference room in Raleigh, five men Jessie had never met came together to consider her future. They were Clifton Craig, the board chairman and state commissioner of public welfare; Jacob Koomen, the state health director; John McKee, representing the commissioner of public health; R.L. Rollins, the superintendent of Dorothea Dix Hospital; and R.S. Weathers, a lawyer from the attorney general's office.

Given the board's record of approving sterilization petitions, Jessie didn't have much of a chance to avoid the operation. As a minor, she had no right to be heard by the board or even to be told about the operation.

The mother had been given a time limit in which to take the necessary steps to bring about her daughter's sterilization. The proper procedure was reviewed which included the information that the petition was to be executed by the Welfare Department and should not be released to the patient.
– Eugenics board minutes, 1956.

Even if there had been someone to speak on Jessie's behalf, it probably wouldn't have helped. A distraught father tried to stop a sterilization for his 13-year-old daughter in 1938, saying that authorities wanted to "butcher her up and experiment when she is innocent." The eugenics board ignored his appeal and issued the order. Thirty years after authorizing the sterilization of that 13-year-old, the board made a unanimous decision to sterilize Jessie. An order went out authorizing Dr. William Bindeman to perform the operation.

'Barren and fruitless'

Several weeks later, Jessie entered Chowan Hospital in Edenton to deliver her son.

At the time, Jessie said, she didn't realize what was about to happen.

"I'd just turned 14, so I didn't know nothing. What was a 14-year-old kid going to know about sterilization and all this crap and being pregnant, you know?"

Someone should have explained what was happening, she said. "Even though I wouldn't have understood it at that age, they could have made an effort to do something, to inform me of something. And then, on top of that, they should have gotten me therapy."

Wright delivered Jessie's baby – a boy – on March 5, 1968. Bindeman, who died a few years ago, performed the sterilization operation several hours later.

Soon after having her baby, Jessie moved to Long Island, N.Y., to live with an aunt, leaving her son to be raised by Woodard.

At 18, still in New York, she married. Her husband wanted children, but she couldn't conceive. She had also been hemorrhaging and having abdominal pains, she said.

After talking with a doctor and one of her sisters, Jessie said, she finally realized that she had been sterilized.

Jessie couldn't understand why it had happened to her.

"Out of all the people in the world," she said. "I was, I am, a good girl, you know?

"I don't think anybody knows how bad it hurts to want to have a kid and you can't," she said. "I used to have a lot of friends, every time they got pregnant, I would not want to be around them. I would just leave them alone. I couldn't stand to be around them when they were pregnant because I wanted a baby so bad."

Her marriage disintegrated. Maggie Skinner, one of Jessie's sisters, said that Jessie's husband was "very cruel about the whole thing.... I remember her (Jessie) crying, telling me that he said she was barren and fruitless," said Skinner, who now lives in North Brunswick, N.J.

Through it all, Jessie was trying to better herself. Even though she had dropped out of school, she earned an associate's degree in human services from New York City Technical College in the early 1970s.

A sister suggested that she talk to the American Civil Liberties Union about what happened to her, and Jessie decided to do so. She eventually filed a lawsuit against the members of the eugenics board, asking for $1 million in damages on the grounds that her constitutional rights had been violated.

Day in court

Her timing seemed right at first. Shortly before Jessie approached the ACLU, the union had in 1973 filed a class-action suit on behalf of another North Carolina woman who had been sterilized, and that case had attracted widespread publicity in a time when issues of women's rights were gaining attention.

The legislature disbanded the eugenics board in 1974, but neither the board nor the state issued any apologies, or even explanations, about its thousands of sterilization cases. The state would continue to release few details about the eugenics program.

When Jessie finally got her day in U.S. District Court in New Bern nine years later, the jurors didn't hear details about the long history of problems at the eugenics board. A wider focus on large-scale abuses at the board might have bolstered her claim, but that evidence wasn't available, even if Judge W. Earl Britt had allowed it.

So Jessie's attorneys focused on the events leading up to her sterilization. "(I)t was not reasonable for the board to rule on that petition without having Elaine come to talk to them, without having any kind of a hearing except looking at those papers and saying, 'Yes, we're going to approve the petition to sterilize,'" said her attorney, Ken Flaxman of Chicago, in his opening statement.

Under questioning from Flaxman and George Daly of Charlotte, former members of the eugenics board and its executive secretary testified they couldn't remember Jessie's case. But after reviewing the records of it, all said they were satisfied that they had made the right decision.

Koomen testified that the sterilizations were a favor of sorts. "The usual response was that we were doing a favor; indeed, we were often asked - not often - we were sometimes asked to sterilize those who had not yet menstruated."

But in 1990 when Koomen talked about his time on the eugenics board, he expressed doubt about the whole program, saying that members of the board "were uncomfortable" in the role.

"Was this the function of the state? Was this a right thing to do? Did we really have all the data at hand? When we were evaluating ... we began to develop a sense, you know, what does an intelligence test mean in this setting?" he told Johanna Schoen, an assistant professor at the University of Iowa, in a previously unpublished interview.

Speaking of Jessie's case and her IQ score of 75, Koomen said "if this had occurred now, we would have let it go."

"And we did it because the law obligated us to. It isn't something we would have volunteered to do - or even suggested," Koomen said.

But Jessie never got to hear even the faintest public statement of regret from Koomen or any other state official.

The jury took just a short time to find that Jessie "was not unlawfully or wrongfully deprived of her right to bear children as the proximate result of any one of the defendants," and her attorneys lost on appeal. They petitioned the U.S. Supreme Court to review the decision but the high court declined.

Jessie was devastated by the failure of the court case, and remained so for years.

"And I didn't know how to escape. So I just crawled into myself. I didn't want to be around people. I didn't want to do anything," she said.

Picking up the pieces

She married again. She got divorced again and said that the inability to have children played a role. She said she went through a period of marijuana abuse and beat it. Now, she lives in a modest apartment and makes it through life as best she can with help from her boyfriend, Calvin Hale, friends, family and her psychiatrist. Her son and sisters take her on cruises and beach weekends.

"I'm spoiled to a point," Jessie said. "I guess they're trying to make up for things that happened to me."

Her son, Tony Riddick, 34, said that the work of the eugenics board was "not far from the thinking of Hitler, when you really think about it. What was happening in Nazi Germany in the '30s and early '40s, you know, that same concept when Hitler tried to make this pure race."

The shoddy science of eugenics made bold predictions that the children of "feebleminded" people would be doomed to failure in life.

Riddick, who still lives in Winfall, earned an associate's degree in the applied science of electronics from DeVry Institute in Atlanta and is the president of his own computer-electronics company.

He wonders whether a sibling of his might have found the cure for cancer or become president. "And this is just my mother, right? But this thing happened to countless people, from what I'm learning," he said. "So did we really do ourselves a disservice by allowing this to happen?"

With the full story of the workings of the eugenics board emerging, Jessie and Riddick want an apology from the state.

Jessie dreams of putting the sterilization behind her, of running a school for abused children.

She also helps a wide circle of friends, visiting shut-ins and ferrying others on errands in her old pickup truck. But occasionally, thoughts about the operation weigh her down, and she stays in her apartment.

"Sometimes, I just don't want to be bothered," she said. "That's when I get depressed."

Tying the Tubes:

Operation has become faster, less painful

By Danielle Deaver
JOURNAL REPORTER

Tubal ligation was considered serious surgery when Elaine Riddick Jessie and thousands of other North Carolina women were sterilized.

Doctors would put a woman under anesthesia and cut one to three incisions into her abdomen. Then a surgeon would reach inside with fingers and instruments to locate her fallopian tubes. The surgeon would cut and "tie" the woman's tubes to prevent eggs from reaching the uterus. This effectively ended her ability to bear children.

Before the 1970s, Jessie and other women who had tubal ligations would have faced several days in the hospital and about a month for a full recovery.

The operation was painful because the surgeon had to cut through several layers of skin and muscle to get to the tubes.

"You'd go home with a sore abdomen but most people would be OK (in a few weeks)," said Dr. Robert Brame, a retired obstetrician/gynecologist who has taught at several medical schools around the state.

Today, women can expect to be out of the hospital the same day they have the surgery and back to their normal routines within a week. The surgery leaves smaller scars and less chance of complications from infection.

Doctors today are also more concerned with the psychological impact of the surgery. They are hesitant to perform surgery on women whose partners don't want it, on young women or on those who have just had children.

"It is important to look at her situation," said Dr. Jeff Deaton, an associate professor of obstetrics and gynecology at Wake Forest University Baptist Medical Center. "You do this to be a permanent procedure with no intent on reversal."

Physically, the surgery has become easier in the past few decades with the advent of laparoscopy, Deaton said.

Laparoscopy is a surgical technique perfected in the past 30 years. Doctors make several small incisions. They put a tiny camera into one and tiny surgical instruments into others. The camera allows them to see inside the area they are operating on without having to make a large incision. Doctors can watch the image on television.

To actually tie the tubes, they usually either burn or cut them, then tie the ends. Some doctors prefer to put a clip or ring around the tubes that also closes them off, Deaton said.

Tubal ligation and the male version of sterilization - vasectomy - have become the most popular methods of birth control in the world. They are permanent and, though they carry some increased risk of medical complications, they are generally considered safe.

People in poorer countries prefer the surgery because it is a relatively inexpensive one-time option for preventing more pregnancies. In the United States and other Western countries, mostly upper- or middle-class people who have already completed their families choose the surgery.

Women who are sterilized also have an increased risk of ectopic pregnancy, a potentially deadly condition where a fetus implants outside of the uterus. They also have risks of complications during surgery, such as an accidental puncture of a bowel or blood vessel.

Vasectomy has traditionally been simpler than female sterilizations. Typically, an incision is made in the man's scrotum. His vas deferens - the tubes that carry the sperm to the penis - are then cut and tied. Patients go home the same day and are usually back to their normal routine within a few days.

The technique has not changed much through the years, though some doctors use a new instrument that allows them to punch a small hole through the scrotum, eliminating the need for stitches.

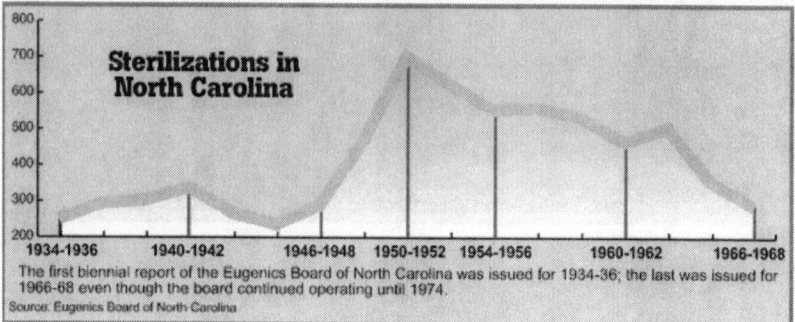

Sterilizations in North Carolina

The first biennial report of the Eugenics Board of North Carolina was issued for 1934-36; the last was issued for 1966-68 even though the board continued operating until 1974.
Source: Eugenics Board of North Carolina

Board did its duty, quietly

Members from five governmental areas
heard case summaries and usually stamped approval

By John Railey and Kevin Begos
JOURNAL REPORTERS

As a new employee of the state attorney general's office in 1960, Chris Coley represented his agency at meetings of the Eugenics Board of North Carolina.

"It was the grunt's work," Coley, 68, said recently.

For 40 years, in a rather routine fashion, the eugenics board made major decisions for more than 7,600 North Carolinians. Five people met in a small conference room in Raleigh and worked through up to 30 cases each month, usually voting to sterilize strangers scattered across the state.

"Later in life, I questioned the fact that I had something to do with sterilizing young women," said Coley, who lives in Raleigh.

Greenhorns such as Coley served on the board, but so did seasoned agency heads. In some years, the board members set the pace; in others, a strong executive secretary did so. But through it all, the board kept quietly working, an empire unto itself, attracting little attention from either governors or average citizens.

As the board neared its end in 1974, members increasingly expressed reservations about its work. But that wasn't the case through most of the board's history. In fact, if there's one common thread that runs through that history, it is the belief of members that they were doing good.

"We thought we were doing a legitimate thing," said Dr. W.A. Robie, 83, of Cary. "I didn't have any idea that anybody was railroading any of them." Robie was a member of the board in the mid-1960s while working for the state Department of Public Health.

The board played its role far from the field, where social workers often pressured women into sterilizations. And by the 1960s, the majority of sterilization petitions were for young black women whom social workers and psychologists often labeled "feebleminded" on the basis of scant evidence.

"This was not the understanding I had of it (the board)," Robie said.

Jacob Koomen of Cary, a former state director of public health, was a eugenics board member from 1966 to 1974. He said that neither governors nor the general public seemed to know much about the workings of the program.

Koomen said that he regularly met with governors Terry Sanford and Dan K. Moore to talk about public-health issues but did not discuss details about the eugenics board. "I would be doubtful they ever knew," Koomen said.

Until two women filed lawsuits against the board in 1973 and 1974, Koomen noted, the press was mostly silent, too. "I don't remember seeing an article about the function of the eugenics board, or someone pointing one out to me. I don't remember a single clipping," he said, even though his agency used a professional service to scan the state's newspapers.

Koomen said he felt that all the decisions he made were justified and that the program was a benefit to many people.

Other former board members echoed his thoughts, saying that the sterilizations the board ordered were beneficial.

"At the time, it was considered good for society," said Gerald White, 76, of Elizabeth City. He served on the board as a representative of the attorney general's office in the early 1950s.

"It protected society and it protected the individuals, as well as their families, who might fall into the role of having to care for a child if the mentally retarded person had a child," he said. "And most of them would be illegitimate."

The records of the eugenics board tell a different story - one of warning signs that were ignored.

In 1955, board member Dr. C.C. Applewhite criticized one of the core foundations of eugenics - that feeblemindedness was inherited. "Suppose this baby soon to be born turns out to be a genius," he said when considering a sterilization petition for a 21-year-old woman. "I have been talking with some of these psychiatrists - some pretty smart boys among them. They have their doubts about this matter of heredity."

White's memory is of a program that benefited society, but another person who represented the attorney general's office during the 1950s said that his boss had doubts.

Attorney General William Rodman was not "too friendly toward the program," wrote board member Worth H. Hester in 1955.

The eugenics board was part of the Department of Public Welfare. Members were not appointed by governors. A state law passed in 1933 set the board membership at one representative each from the attorney general's office, Dorothea Dix Hospital, the state Department of Public Welfare, the state Department of Public Health and the state Department of Mental Health.

Sometimes the directors of these organizations served on the board themselves. Other department heads sent an ever-rotating round of junior staff members to represent their agencies.

"I was new and naive and hadn't done anything like that," said Robie, the doctor from Cary.

State employees sent to the meetings didn't get crash courses in child behavior or the effects of environment on IQ before joining. But at meetings that lasted at least two hours, board members would take up petitions for sterilization that touched on a wide range of other issues, albeit briefly.

The board often voted to sterilize people who were mildly retarded, as well as some people of normal intelligence who were classified as feebleminded because of flawed IQ tests. Results of IQ tests were included in sterilization petitions that came from county departments of public welfare across the state. As the board's executive secretary got these petitions, she would condense them to one-paragraph summaries for the board's consideration.

While the full file on each case was available in the board's meeting room, members have said they sometimes voted on the basis of the summary alone. Though evidence of some dissent among board members is plain in some board records reviewed by the *Winston-Salem Journal*, several former members said the process was smooth.

They emphasized that they didn't always vote to sterilize. "I remember I thought there were some better off sterilized that didn't pass," Robie said. "It wasn't a carte blanche thing."

But sterilization was ordered in more than 90 percent of cases. A network of surgeons across the state, usually in private practice, carried out the board's orders.

Ralph Potter of Fayetteville, who served on the board in the early 1960s while with the attorney general's office, noted that the board functioned in a different era.

"By the nature of the times, it probably didn't draw the attention that it probably deserved," said Potter, 65.

Read This: Records unexpectedly available

By Kevin Begos
JOURNAL REPORTER

Officially, the records of the Eugenics Board of North Carolina are still under seal at the N.C. State Archives in Raleigh. No one has ever been given full access to the records, though some researchers have recently been given small selections.

Around the county, "the history of things has been to put up roadblocks," said Paul Lombardo, the head of the Center for Biomedical Ethics at the University of Virginia. "(Officials) make it as difficult as possible for people to find this stuff."

The *Winston-Salem Journal* was given access to thousands of pages of the North Carolina records by Johanna Schoen, an assistant professor of women's history at the University of Iowa.

Schoen was a graduate student at the University of North Carolina at Chapel Hill in the late 1980s when she got permission from the attorney general's office to review the records. She was rebuffed at first by staffers at the archives but continued to work on the small amount of material open to the public.

One day a staff member put a roll of microfilm in her hands. Schoen isn't sure why, but believes that the archives staff wanted the truth about the eugenics board to be known. She was allowed to copy thousands of pages of documents.

Schoen gave the Journal access to the records on the condition that the medical privacy of the people who were considered for sterilization be honored.

The archives denied an oral request by the Journal in August for access to the eugenics board records.

Fewer People: Peru resorting to sterilization?

By Kevin Begos
JOURNAL REPORTER

For many poor women in Peru, a large-scale sterilization program is not a thing of the past. Fernando Carbone, Peru's health minister, said this summer that as many as 200,000 women may have been forced or bribed into having sterilizations from 1996 to 2000.

Most of the victims were members of the Quechua and Aymara tribes from remote areas in the Andes.

Christina Ewig, who researched public health issues in Peru last summer, said that the politics of the sterilization scandal has many layers. Ewig just received her Ph.D. in political science from the University of North Carolina at Chapel Hill, and is now teaching at the University of Wisconsin.

"There is evidence of some coercion or at least uninformed consent prior to sterilizations. I highly doubt that the numbers are anything near what the current government claims," she said.

What is clear is that former president Alberto Fujimori and members of the Peruvian military wanted to slow population growth by using sterilization and that some health-care workers had quotas to fill, she said.

"It was very important to Fujimori himself - there were indeed goals for numbers of sterilizations - that's been documented," Ewig said.

While there were abuses in the sterilization program, Ewig thinks that Carbone exaggerated the number to discredit birth control programs because of his conservative religious beliefs.

An official investigation into the sterilization program by the Peruvian Congress concluded that there were "massive, compulsory and violent violations of fundamental human rights," and suggested that Fujimori, who is living in exile, might be guilty of genocide under international law.

Forsyth in the Forefront
Medical school to probe its role in county plan for sterilization

By Danielle Deaver
JOURNAL REPORTER

As World War II wound down in 1944, Dr. C. Nash Herndon of the Bowman Gray School of Medicine described a eugenic sterilization program in Forsyth County using language eerily reminiscent of Nazi experimentation.

"In September 1943, a project aimed at eugenic improvement of the population of Forsyth County was begun in co-operation with Dr. J. Roy Hege, Forsyth County Health Officer. This project consists of a gradual, but systematic effort to eliminate certain genetically unfit strains from the local population. About thirty operations for sterilization have been performed," Herndon wrote in his annual report for the Department of Medical Genetics for the 1943-44 school year.

Sterilization operations not approved by the Eugenics Board of North Carolina were illegal until 1963. Though the state generally ignored sterilizations performed by private doctors on willing patients, cases of eugenic sterilization such as those described by Herndon had to be approved by the board.

Officials with the Wake Forest University School of Medicine today condemn the eugenic research.

"I think that the concepts and the practice of eugenics is wrong and unethical and would in no way be approved or condoned in modern medical times," said Dr. William B. Applegate, the dean of the medical school.

Applegate said that the medical school will set up a review committee of faculty and administrators to look into the matter. "It is our duty to the public to be squeaky clean," he said.

A brand new thing

When the Bowman Gray School of Medicine opened the country's first department of medical genetics in 1941, no one in the department other than its new director, Dr. William Allan, had any idea how important genetics could become.

But Allan viewed genetics as something quite different from what most people hope for today. He was an early proponent of eugenics - the movement to keep people considered unworthy from reproducing.

"It seems to me that the only way to attain the goal of positive eugenics is to actually practice negative eugenics – the prevention of the birth of the mentally and physically unfit," he wrote in a paper presented to the Eugenics Research Association in New York in 1936.

Bowman Gray is now part of Wake Forest University Baptist Medical Center, the city's largest employer and one of the most respected academic medical centers in the country. From its beginnings in genetics, it has gone on to establish a Center for Human Genomics, where cutting-edge research has been done since its creation in 2000.

But for the first few decades it was in Winston-Salem, the school's researchers enthusiastically pursued a negative form of eugenics - as evidenced by Herndon's papers found by the *Winston-Salem Journal* in the Dorothy Carpenter Medical Archives at the Wake Forest University School of Medicine.

Herndon's description of the Forsyth eugenics program as a "systematic effort to eliminate certain genetically unfit strains" sounds like "negative eugenics in it purest form," said Johanna Schoen, an assistant professor of women's history at the University of Iowa who researched the eugenics board while studying at the University of North Carolina at Chapel Hill.

According to Herndon's report, the Forsyth program was a cooperative effort between local elected officials and the Department of Medical Genetics at Bowman Gray.

"The expense of this project has been borne by the Forsyth County Commissioners and necessary operations have been performed at the Forsyth County Hospital. Genetic work-ups and medical affidavits have been supplied by this department," Herndon wrote.

"It (the Forsyth program) seems like a usurpation of authority that is just amazing to me," said Schoen. "Partly because if this is the kind of sterilization they wanted to do there was already a venue to do them through the eugenics board."

No mention of the Forsyth program can be found in the minutes of the Forsyth County Board of Commissioners in 1943 or 1944. County health-department records for those years have been lost or destroyed. At the request of the Journal, county officials are searching for any records pertaining to the program.

According to minutes of the Human Betterment League of North Carolina recorded in November 1949, Herndon "had himself performed six operations in the past week and told of the very advanced policy of Baptist Hospital." It is unclear whether those sterilizations were part of the Forsyth program or ordered by the state eugenics board.

Still, understanding how county officials and Bowman Gray teamed up to create a county-level eugenics program that apparently operated outside the scope of the state eugenics board is complicated. Part of the answer may lie in eugenics board records sealed by the state; parts may yet be hidden in the medical archives at Wake Forest.

"It appears to one there could have been some association between the Bowman Gray School of Medicine and the county," Applegate said. "I believe, from the documents, that our department of medical genetics was somehow involved."

Whatever the answer, any examination of the program and the medical genetics department has to begin with William Allan.

A stark beginning

Allan was one of the pioneers in genetics and he was possibly one of the most visionary members of the early community. He wrote his first study in 1916, about the genetics of migraines; by the end of his life, he had written or co-written 93 papers.

He was in private practice until 1941, when he moved from Charlotte to Winston-Salem to start the medical genetics department at Bowman Gray.

Allan was convinced that hereditary diseases could be conquered by prevention, much like some infectious diseases had been.

He could even see a day when genes could be manipulated to eliminate harmful attributes. "Numerically the bulk of our hereditary diseases comes on too late in life to be used in any negative eugenics program as understood at present; maybe some way to modify genes might be found some day," he wrote in 1940.

Much of his work was based on field studies, including statewide surveys of people with muscular dystrophy and other crippling diseases, and surveys of 800 people in 15 counties who were blind.

"These surveys were designed to obtain data on the incidence and genetic mechanisms of the diseases in question, data that are essential for the formulation of a practical program for reducing the supply of grossly defective children in our population. The Department of Medical Genetics at the Bowman Gray School of Medicine is now engaged in a continuation of this work," according to a tribute to Allan written by Herndon in 1944.

Allan's dream was a statewide bank of genetic information. "Armies" of health workers could be mobilized to catalog genetic backgrounds of prospective parents, he said. People who wanted to get married could pick up a license to have children at the same time, if they were genetically fit.

Allan thought that at some point much of the work done by eugenicists and the state would be unnecessary.

"Will the members of families with hereditarily blind, deaf or crippled children take eugenic advice? ... After some years experience in practicing negative eugenics among the members of afflicted families, the writer knows that they are eager for reliable information about their family disability and will govern their lives accordingly," Allan wrote in the late 1930s.

Allan knew how desperate some people were to prevent hereditary diseases from being passed down. During his lifetime, he answered hundreds of letters from people with questions about genetics. Some wanted to know if they could marry first or third cousins; others wanted to know if asthma should prevent them from having a child.

The letters came from as close by as Greensboro and as far away as New Mexico, and Allan answered them all.

Usually, he told people that they should go ahead and have children.

The pioneering nature of Allan's work in genetics began to be recognized after his death. The William Allan Memorial Award for Outstanding Work in Human Genetics is still given by the American Society of Human Genetics. But Allan never got to push his ideas. He died in 1943.

He was succeeded by Herndon, a protege equally excited by the possibility of improving the human race.

A new breed

Under Herndon, the department continued to conduct surveys of blind, deaf and crippled people, looking for the links of hereditary disease.

But members of the department also continued to research - and possibly implement – eugenic ideas. Herndon was an enthusiastic supporter of eugenics. He was the president of the American Eugenics Society from 1953 to 1955 and a president of the Human Betterment League of North Carolina, a group that had helped to energize the sterilization movement after World War II.

In an updated paper called "Suggested Programs of Research in Medical Genetics," Herndon implied that eugenics could go far with the Bowman Gray Department of Medical Genetics behind it.

North Carolina was the perfect place to try out a genetics program, Herndon wrote, because so many generations of each family lived in the state, making it easier to track family trees and the diseases that flowed through them.

A more subtle reason was in play, too. Lack of protest from the community would make it easier to proceed on what could be a controversial mission, he wrote.

In the same paper, Herndon starkly laid out his case for eugenics when he wrote about how to eliminate Huntington's disease - a degenerative illness that affects a person's ability to walk, talk and think. It strikes people in their '30s and '40s and can lead to mood disorders and depression and eventually causes death.

The disease was discovered in 1872 and quickly became a target for eugenicists because it seemed easy to figure out - every child of a Huntington's patient had a 50 percent chance of contracting the disease. Doctors thought that a single gene caused the disease. Since there was no way to test for the gene in those days, the only way to know if people had the disease was to wait and see if they developed symptoms.

But many people would have already had children by that age, potentially adding another generation of people with Huntington's. That left only one eugenic option, Herndon pointed out in his report.

"Sterilization of all children of affected persons before the childbearing age would result in elimination of the defective gene.... Such a program would involve a 'wastage' of approximately 50 percent, in that only half the persons sterilized would later develop (Huntington's), but such a wastage would seem to be justified in view of the benefit to the general population that would result," he wrote.

It would have been nearly impossible to actually carry out such a plan, said Dr. Francis Walker, who is now a professor of neurology at the school of medicine.

"It's really hard to sterilize someone who has a 50 percent chance of getting the disease. That's kind of harsh," he said.

Even without the ethical considerations, the plan would not have worked, Walker said.

"You could never cut out this disease completely because there are always people whose (mutations) are sneaking up," he said. "About 5 percent of our patients we could never track it down completely in families."

Bad science?

The team's basic mistake about the origins of Huntington's casts doubt on how much its members really knew about genetics and the diseases that they were trying to eliminate.

Some of the medical knowledge about genetics was pretty good for the time, said Garland Allen, a professor of biology at Washington University in St.

Louis who has studied the history of biology and eugenics.

"In terms of the general transmission of (simple genes that carry characteristics such as eye color), transmission was quite well understood by the 1940s and '50s," Allen said. But scientists did not know how likely people were to show the characteristics of the gene if they inherited it.

"It was assumed that if a gene was transferred from parent to child, it would be automatically inherited. And that wasn't the case," he said.

There was no way that William Allan or Herndon could have known how much about genetics would be discovered during the next 60 years. But they did seem to realize that they didn't have a good understanding of some aspects of genetics.

In his papers and letters, Allan emphasized the importance of the "pedigrees," or family trees, that he and his assistants drew.

"Can you imagine a eugenic measuring rod for parents that does not start with a study of their pedigrees?" he wrote.

Today, doctors know that while heredity plays a role in some diseases, it usually does not determine whether people actually get that disease. Few would recommend that a patient forgo having children based only on family history. And new methods of genetic profiling make it much easier to find out if someone is carrying a gene for a disease.

Recessive genes caused some of the diseases that Allan and Herndon were researching, including muscular dystrophy. Herndon knew that many diseases were caused by recessive genes but did not yet know how they work.

"At the present time we do not have sufficient data concerning the frequency and distribution of such genes to recommend any program of control which would show much promise of any really significant reduction in the incidence of such diseases even after several generations," Herndon wrote.

Herndon also knew that environment plays a large role in the development of some diseases and that environmental factors help explain why some people with a family history of conditions such as diabetes never develop them. But despite his knowledge of his own shortcomings and of the unknown role that environment plays in the disease process, Herndon continued to go forward with eugenic research and genetic counseling.

It was typical of the geneticists of the time, who thought that they could apply the genetics that governed simple attributes like eye color to more complex psychological behaviors, said Garland Allen.

"The eugenicists were mostly interested in complex mental traits like feeblemindedness, alcoholism, and there was no evidence at all that those were really inherited. So there was a lot of confusion," he said.

The early genetic researchers were using faulty science, said Applegate, the dean of the medical school.

"There were certain genetic assumptions made then that we know now were wrong," he said.

A program marches on

The department's involvement in eugenics continued at least through the early 1950s. After Herndon's first mention of the Forsyth Eugenics Project in his annual report for 1943-44, department members continued their work on the project at least through 1948. Herndon's report for 1947-48 says that "steady progress" was being made and that the program was continuing and going well.

After that, the Department of Medical Genetics stopped doing its own reports until 1952, when there is no mention of the Forsyth project.

Members of the department also continued to study the genetics of blindness and certain diseases, including muscular dystrophy.

Researchers tried to use the information that they had to offer free genetic counseling. Herndon mentioned in his annual reports that the doctors counseled people who came to the hospital with questions about hereditary diseases and whether it was safe to marry relatives.

But involuntary sterilization was still an issue during the 1950s, and it involved members of other departments at Bowman Gray.

Dr. Frank Lock, a professor of obstetrics and gynecology and later head of the department, and Dr. James F. Connelly, also a doctor at Bowman Gray, co-wrote a paper for the Bulletin of the American College of Surgeons in 1953 about the indications for the sterilization of women that included a section on involuntary sterilization.

"North Carolina is considered to have one of the best eugenics laws in the nation, and the number of individuals sterilized under this law is high in comparison with the rest of the nation. (Moya) Woodside has pointed out, however, that the number of sterilizations performed under this law has probably not been high enough," the doctors wrote.

In 1950, Woodside published Sterilization in North Carolina, a book that examined the state's sterilization program. The doctors also defended the state's program. "The individual's own rights are clearly protected by due legal processes, and there is a specific section of the law which relieves the surgeon of any responsibility for performing the operation."

In the end

Late in his life, Herndon seemed to deny what had gone on in the medical-genetics department under his watch.

He spoke in 1976 with Elizabeth Allan Berger, the daughter of William Allan, as part of a series of oral history interviews set up by the Carpenter Archives. During the interview, Herndon denied that the department had ever engaged in "negative eugenics" - preventing people from having children by sterilizing them.

"From a semantic viewpoint, the word 'eugenics' acquired something of a negative connotation I think... during... um, the, well, right after the Hitler period and so on," Herndon said.

"Yes. And all the compulsory sterilization over there and that sort of thing," Berger said.

"Which was not at all what these people were interested in talking about. We used to talk about positive eugenics," Herndon said.

But the papers that Herndon wrote and the work that Bowman Gray researchers do show that the department engaged in negative eugenics through at least the 1950s - not the least of which was the Forsyth eugenics program.

Applegate said he plans to make the results of the medical school's investigation public.

"This is a worrisome issue, and to the extent that our school was involved in it, I would like to see the details," Applegate said. "As best we can, we need to understand the past to guide the future and we need to be as transparent as possible."

Part of the present and future is work being done by the section of medical genetics, which is part of the pediatrics department at the school of medicine.

The department that Allan founded still works with children with birth defects, but in a much different way. Genetic counselors in the department talk with women who come in for prenatal testing, educating them about potential problems, and advising them about their options.

Medical geneticists in the department diagnose genetic disorders and birth defects in babies. The doctors also act as case managers for children. They see the patients every year for follow-ups to tell them about new treatments, and they coordinate care with other doctors and suggest support services, said Dr. Tamison Jewett, an associate professor in the department of pediatrics, division of medical genetics, and director of clinical services in genetics.

Like their founder, present-day members of the department want to alleviate the suffering caused by these diseases. But unlike Allan, they have the benefit of knowing that many birth defects can only be treated, not eliminated.

"If you're good at this kind of thing, which I think we are, then you can really do so much good, helping people through these times because birth defects aren't going to go away," Jewett said. "We'd like to think they are, but genes are always changing."

Sign this or else...

A young woman made a hard choice, and life has not been peaceful since

By Kevin Begos and John Railey
JOURNAL REPORTERS

ATLANTA– Nial Cox Ramirez remembers every detail of what happened to her in 1965, even though she has been trying hard to forget.

Ramirez had a choice to make, and it was a wrenching decision for an an 18-year-old who had just had her first child.

Her options? Sign a form from the Eugenics Board of North Carolina "consenting" to be sterilized, or have welfare payments for her mother and six brothers and sisters cut off.

"To sit and think about it literally eats you - slow, real slow. It eats you piece by piece," Ramirez said recently. "That's why I don't want to go back there. Because it's a hell within a hell that you going through. It's like a cancer that eats."

Ramirez, now 56, lives outside Atlanta. She is still trying to make sense of what happened.

"Why me? That's what I want to know. Why me? Why you want to bother me?" she asks.

In the thousands of pages of sealed records of the eugenics board reviewed by the *Winston-Salem Journal*, some of the answers that Ramirez has been searching for can be found.

On July 1, 1961, Sue Casebolt took over as the executive secretary of the eugenics board. Before the month was over, she was pushing a new agenda that would target girls like Ramirez.

I now propose to have as my objective as Executive Secretary to work to promote earlier use of the (sterilization) program; that is, after the first rather than third of (sic) fourth child, which would result in prevention of problems requiring staff time, money, and use of other needed community resources. To this I plan to use all resources available to secure information as to persons who need to be offered the service. A few of these are: Mental Health Clinics. 2. County Health Officers. 3. Public Welfare records such as APTD and ADC.

— Eugenics board minutes, 1961.

Emotional blackmail

In her tidy home, Ramirez opened up and talked at length about her life. She was warm and friendly—until she started talking about the sterilization.

"I tried to bury it. I tried to get rid of it. I tried to forget all about it," she said. "But it comes right back fresh, just like it was yesterday."

It was 1964, and Gov. Terry Sanford and President Lyndon Johnson were pushing their visions of a new South.

But the word hadn't gotten to Plymouth, the river town in Eastern North Carolina where Ramirez grew up. It hadn't gotten to the housing project where she lived with her mother and six brothers and sisters.

After Ramirez got pregnant by her boyfriend, a white woman from the Washington County Department of Public Welfare started making more frequent visits.

"She came almost every other day, carrying her little pocketbook and her little briefcase," said Ramirez, referring to social worker Shelton Owens Howland. "And she goes all into details. Every little detail. She would always tell me, 'Your family is going to starve because of what you did. If you don't do this, we going to take this check away from (your mother)'."

"I was so disgusted with her," Ramirez said. "And really, if I knew how to have a gun, I would have shot her. Because it doesn't make any sense."

It didn't make any sense to Ramirez, but what happened to her was hardly an accident.

I plan a tickler file on all persons whose names reach me regardless of age in order that they may be picked up as they reach the child bearing age.
— Casebolt, Eugenics board minutes, July 27, 1961

But Casebolt was only part of a chorus that had been calling for a stronger sterilization program. And for years, there had been suggestions of targeting the black community. In a 1950 book on the North Carolina eugenics program, it was taken as fact that poor blacks had more mental problems than other groups.

There is need for special education among the lower-class Negro groups, since it is here that fertility is highest and mental defect more prevalent.
— *Sterilization in North Carolina*, Moya Woodside,
University of North Carolina Press

The sterilization program had been conceived in 1929 as a tool to be used in state institutions, and for the first 20 years, the majority of sterilizations took place in state hospitals and schools for the disabled and the delinquent.

But as members of the mental-health community began to wake up to the fundamental flaws of the sterilization program, the state Department of Public Welfare turned out to be an even more enthusiastic partner. North Carolina was the only state in the nation where social workers had the power to initiate sterilizations in the general population.

... the Honorable David Henderson, representative from Mecklenburg County, wished the board to consider the possibility of expanding the provisions of the eugenics legislation. He wishes consideration to be given to the possible sterilization of families, who, while receiving financial assistance continue to have more children. — Eugenics board minutes, 1951

Officially, the final decision on sterilization petitions in North Carolina rested with the five members of the eugenics board. In practice, however, social workers at the county level had tremendous power - and that was obvious to Ramirez.

"I don't know if she (Howland) hated black people or what," Ramirez said. "But she had this attitude, this nasty way of talking. Like you're nobody, and she's somebody. She (was) God, and I'm a little rat running around on the floor."

In the sterilization petition, Howland wrote that Ramirez was argumentative and lazy and had been told that the welfare department "would not assume the responsibility of supporting all the children she would bring into this world.... Our agency is thoroughly convinced that the only way to keep a family of this type from reproducing itself is to rely on sterilization."

Howland, who now lives outside the state, declined repeated requests to comment for this story.

Comments from Elsie Davis, a Fayetteville social worker who was working in the 1960s, echo Ramirez's impression that many social workers carried an inherent bias.

"The expectation was that black people were not able to take care of themselves," Davis said in a previously unpublished interview done in 1989 by Johanna Schoen, an assistant professor at the University of Iowa. "They were all illiterate, retarded. So it was consensus that these women don't have any rights. So we can say to them that they can't have any children."

"It was a system rather than the individual, who didn't have any rights at all," Davis said, and records from the eugenics board are filled with examples of the bureaucracy that she spoke of.

On August 23, I visited the Bladen County Welfare Department.... The Health Officer had tabulated all out of wedlock pregnancies for 1962, and indicated an interest in having the younger mothers evaluated ... and it was felt that to begin with the younger ones who have a longer reproductive period would be helpful to all concerned. — Eugenics board minutes, 1962

Pressure from all sides

Ramirez gave birth to a daughter, Deborah, in November 1964. The pressure from the welfare department continued.

"There was another lady up in the courthouse on the second floor," Ramirez said. "(She said) 'That's good for you. That's good. You should have that (sterilization). You shouldn't have any more kids.'"

Trying to cope with a newborn child and the poverty of her family, she had to make her choice.

"And what am I supposed to do?" she said, her voice cracking. "Why should my family - my sisters and brothers - starve for something I did?"

With the memory came tears.

Soon, a final petition for sterilization was sent to the eugenics board. Ramirez's life was summarized in one paragraph that was almost all opinions, not medical or psychological fact. It read, in part:

Nial Ruth usually runs errands and buys the groceries but takes no responsibility about the house. She has worked at field work but becomes quite argumentative and thinks she is never paid enough. She does not get along well with her siblings.

On Feb. 10, 1965 - three months after Deborah was born - Ramirez was sterilized by Dr. A.M. Stanton at the Washington County Hospital in Plymouth. Ramirez said she asked Stanton not to do the operation but report that he had. Stanton told her that he couldn't do that, she said.

After the fact, Ramirez said, some people quietly questioned the decision to perform the operation.

"The nurses was nice. Some of the nurses was good, and they would say, 'We're sorry this is happening to you.' Some of the nurses even said (it was wrong.)

"They may have it in their heart that it's not right, but they're not going to walk up there and say, 'Doctor, what you're doing is not right.' They whisper it to you, but they just saying it to you. They don't want nobody else to know that they saying anything, anything at all about it."

The black community wasn't any help either, she said.

While recovering in the hospital, Ramirez said, she thought that Stanton hadn't just been told to do the operation - he wanted to do it.

"I used to have nightmares of that stupid doctor, that stupid caseworker come walking with her case ... it's like the devil coming with the pitchfork. But thank God I don't have those dreams no more. That's one thing God took away."

Stanton, who still lives just outside Plymouth, said recently that he did the sterilization because he was asked to do so. "I just complied with the eugenics board, that's all," he said.

Getting away

In the late '60s, Ramirez moved to New York. She worked at Hempstead General Hospital for 12 years as a nurse's aide, she said, until she broke her ankle and had to go on disability. "I took care of my daughter, I sent money home to my mother," Ramirez said. "So how can a crazy person hold a job? And I worked in a hospital with sick people."

Deborah lived with her grandmother at first, but Ramirez soon brought her to New York.

"Everybody was proud, tell me what a good mother I am. How I take good care of my child. Crazy people don't do that. Retarded people don't do that," she said.

Deborah Chesson, 38, lives with her mother, and the two look after each other's needs. Deborah graduated from Elizabeth City State University, and now works for a computer company.

After living in New York for a few years, Ramirez in 1973 became the first woman to file a lawsuit against North Carolina's eugenics board and the social workers and doctors who supported it, charging that the sterilization had violated her constitutional rights.

The early publicity about the lawsuit contributed to a decision by the legislature to disband the eugenics board, but Ramirez lost the case in the end.

Ramirez finds comfort in her faith, but even with prayer and reflection she still wants North Carolina to apologize for its sterilization program.

"What they did to me was wrong," she said. "It was really, really, wrong."

Ramirez has been one of the few to speak out about what happened to her, but in the records of the eugenics board there are more suggestions of a system that had gone wrong for many black women, and for whites, too. Instead of in-depth discussions about individual cases, there are phrases that read like gossip, not science.

Because her family is quite casual in it's (sic) approach to life, the welfare department feels that she should have the sterilization operation before she is allowed to leave the Murdoch School permanently.
    — Eugenics board petition for a 17-year-old white woman, 1962

There is indication that this girl will continue to be sexually promiscuous. She is not capable of giving children the care they need.
    — Eugenics board petition for an 18-year-old black woman, 1958

Stanton said that cases had to be rated individually, but that the sterilization program "was probably a good thing.... I think some people did it on purpose (had children) to get a little bit of extra money from the welfare department."

Ramirez wants to forgive the eugenics board and Stanton, perhaps because she believes that it will help hasten her own healing.

"And Doctor Stanton, wherever you are - I hope God blesses you," she said. "In your heart of hearts you know what you did was wrong. Think about it. Would you like it to be done to your daughters? You was an M.D. You don't got to do what the social workers tell you. Doctor, you went to school to save lives.... And you're the reason that I can't trust doctors."

"If I was a doctor, I would say 'No, I'm not going to do that. This is a young girl – she's just growing up. Why would I do that to her?'"

Benefactor With a Racist Bent

Wealthy recluse apparently liked the looks and potential of Bowman Gray's new medical-genetics department

By Kevin Begos
JOURNAL REPORTER

The official version of the story from Wake Forest University has long been that a "New York philanthropist with a deep interest in population genetics" made a $100,000 gift of stock to the fledgling Department of Medical Genetics in 1953 – the equivalent of more than $650,000 today.

The whole story is this: The philanthropist was Wickliffe Draper, a wealthy recluse with a record of paying for research that tried to "prove" that whites were superior to blacks. The Bowman Gray School of Medicine got the money, but as part of the deal one of the most controversial patrons of racist science got his identity hidden.

"There appears to be some association with Draper. We don't know what the total amount was. We don't know specifically what it was used for," Dr. William B. Applegate, the dean of Wake Forest University School of Medicine, said last week.

Scholars taking hard looks at Draper say that his visit to a Nazi eugenics conference, his bankrolling of the racist tract "White America" and support for sending black Americans back to Africa are part of a pattern – the money promoted the now-discredited eugenics movement and radical racial politics.

"Draper wished to use science as a way to stop the civil-rights movement," said William Tucker, a professor of psychology at Rutgers University and author of *The Funding of Scientific Racism:Wickliffe Draper and the Pioneer Fund*, published this year.

Applegate said he couldn't condone the school's acceptance of money from Draper, but didn't want to judge his predecessors.

"Clearly the Bowman Gray School of Medicine took a gift from a source in those days that we would not take from that source today, for ethical reasons," said Applegate, who is creating a faculty committee to investigate Draper's ties to the school. "It would be impossible to accept money from a source with those views at this school today. It's absolutely, totally unacceptable."

The official history of the school of medicine is *The Miracle on Hawthorne Hill*, by Dr. Manson Meads. It refers to an anonymous patron and a gift of $100,000 to be spread over 30 years to pay for the first professorship of medical genetics at the school.

It was 1949 when Draper met Dr. C. Nash Herndon of Bowman Gray. Little is known about their meetings, but Herndon was playing a major role in the expansion of the North Carolina eugenic sterilization program at the time.

Research by the *Winston-Salem Journal* shows that in 1951, Dr. C.C. Carpenter, the dean of the medical school, sent Herndon a memo about grants. The school policy was to assume responsibility if the other source terminated, Carpenter wrote, and "this would be true in your case in regard to payment of your salary through the Draper grant."

Paul Lombardo, the head of the University of Virginia's Center for Biomedical Ethics, said that before Draper gave money to Bowman Gray, a similar offer to another school had raised ethical questions and fallen through.

In 1948 Draper suggested a gift of $100,000 to officials at the Dight Institute at the University of Minnesota to begin a "human genetics project," but director Shelton Reed wrote in a letter that year that Dight probably wouldn't get the money.

"Colonel Draper has very definite ideas as to what the subject of human genetics encompasses," Reed wrote, adding that meant "improvement of the American people by shipping the Negro inhabitants back to Africa. My remark about Colonel Draper is not flattering, but I think you will agree it is generally correct."

Born in Massachusetts in 1891, Draper had old Kentucky blood on one side of the family, old Puritan on the other, and was sure that kind of family tree represented the true American. He inherited a multimillion-dollar textile fortune in the early 1920s and never worked at a job other than military service in World War I and World War II. A lifelong bachelor described as extremely reclusive by even his admirers, Draper lived in a huge New York apartment filled with stuffed big-game trophies he'd shot on trips to exotic parts of the world.

The $100,000 gift to Wake Forest was one of Draper's largest, Lombardo said, and an examination of what kind of science he was interested in raises questions about what he expected in return for his money.

In 1935 Draper attended the Nazi's International Congress for the Scientific Investigation of Population Problems in Berlin, Lombardo said. The Honorary Chairman of the meeting, Wilhelm Frick, was later convicted during the Nuremberg war-crimes tribunal and hanged in 1946.

Draper's companion at the conference, Dr. Clarence C. Campbell, gave a speech that resounded across the Atlantic. "The difference between the Jew and the Aryan is as unsurmountable (sic) as that between black and white," and closed with a rousing "To that great leader, Adolf Hitler!" Time magazine noted.

The next year Draper chose Campbell as a judge for an essay contest he funded, and in 1937 Draper and eugenicists Harry Laughlin and Frederick Osborn launched the Pioneer Fund as a vehicle for distributing grants. Tucker notes that a proposed Pioneer budget for 1937 mentions "two German films referred to by Colonel Draper have been received," and that one, The Hereditary Defective, was shown at 28 U.S. high schools through Laughlin's efforts. Draper continued to work with Laughlin to find ways to use his money to further science that focused on racial issues. Lombardo said there is conclusive evidence that Draper funded a special printing of Earnest Sevier Cox's racist tract "White America" that was sent to every member of Congress in 1937.

Tucker has shown that Draper gave $215,000 to the Mississippi Sovereignty Commission in the 1960s, a group that used the money for a failed effort to stop the passage of the 1964 Civil Rights Act.

In the late 1950s Kinston, N.C., native Harry Weyher became the president of the Pioneer Fund, a position he held until his death this year.

Through Weyher, Draper gave $500 to Wesley Critz George, a staunch advocate of segregation and a professor at the University of North Carolina, to help distribute a 1961 anti-integration pamphlet. Draper sent George $1,000 Christmas checks every year until 1972, Tucker notes.

Pioneer Fund ties to North Carolina expanded in later years, as Marion Parrott of Kinston and Tom Ellis, a Raleigh lawyer, served on Pioneer's board of directors. Ellis' Coalition for Freedom, a part of the Jesse Helms political machine, received $195,000 in Pioneer grants in the 1980s, the equivalent of more than $300,000 today. Ellis did not return calls seeking comment.

Pioneer Fund grants paid for much of the research that supported The Bell Curve, a 1994 best-seller that argued that blacks are genetically inclined to be less intelligent than whites.

J. Philippe Rushton, the current Pioneer president, said there's no doubt some of Draper's past is a public-relations embarrassment, but "that has absolutely nothing to do with the Pioneer Fund."

Because the Pioneer Fund doesn't take stands on political issues, Rushton said it makes no sense to apologize for Draper's past.

"We wouldn't condemn Draper's views on segregation or somebody else's views on integration," he said. Those who attack Draper and Pioneer downplay details that don't fit allegations of racism, Rushton said.

One of the original members of the Pioneer Board of Directors was John Marshall Harlan, who later became a Supreme Court justice and voted for the landmark 1954 Brown vs. Board of Education school-desegregation ruling.

Draper's critics reject those defenses. "(He) made a point of funding people who shared his views on race, and he also made a point of trying to hide what he did," Lombardo said.

Draper was extremely secretive, but even his scant correspondence suggests that he must have believed that Herndon shared his general views on race. Tucker quotes from a 1954 letter to Osborn:

"I should be reluctant to assist investigators whose personalities and view points were markedly alien to my own," Draper wrote, adding that he believed in "measures to promote considerable ethnic homogeneity...."

Rushton said he knew nothing about Draper's gift to Bowman Gray, or Herndon's role in the expanded eugenic sterilization program in North Carolina that would become dramatically more racist in the 1950s and 1960s.

"Maybe that's good or bad charity, depending on your values," he said of possible links between Draper's gift and sterilizations in North Carolina.

Applegate is just beginning to look into ties between the school, Draper and sterilizations in North Carolina, but he rejected the entire concept of eugenics.

"To me, the whole concept of involuntary sterilization sends a chill down my spine," Applegate said. "I just think that it's morally wrong. The very concept of that is profoundly upsetting to me and to the leadership and the faculty of the school."

Comes a stranger
Geneticist combed Watauga, creating, studying family trees

By Danielle Deaver
JOURNAL REPORTER

When Dr. William Allan paid personal visits to Watauga County residents in the early 1940s, he wanted to hear family stories. He wanted to know who had died and who was still living. He wanted to know what diseases ran through the generations and whether any strange features such as six-fingered hands may have recurred in the family tree. He wanted to know if any family members were "feebleminded."

Though it may have seemed odd that the head of the Department of Medical Genetics at the Bowman Gray School of Medicine would drop by hundreds of mountain homes, his visits were far from social. Allan was studying genetics in Watauga County – collecting the family trees of residents. His research would become the basis for much of the science that he and others did at Bowman Gray School of Medicine.

Allan spent much of his time in Watauga County visiting family reunions with an assistant. At every occasion, he would pull out a pad of paper and a pencil and – with the help of older family members - he would start drawing family trees.

The family trees are still at the Dorothy Carpenter Medical Archives at Wake Forest University School of Medicine. Five boxes overflow with hundreds of folders with family trees drawn by hand on draft paper.

Allan was attempting to trace the heredity of disease – serious conditions such as blindness and smaller defects such as extra fingers or toes that ran through families. He thought that his work would contribute to the small but growing body of knowledge about heredity.

He picked Watauga County for its small, multi-generational population. There had not been much in the way of immigration to the area, so the genetic lines had remained fairly pure, Allan explained in letters. He wanted to collect family trees - or pedigrees, as he called them - for all 18,000 people in the county. He nearly reached his goal, collecting about 75 percent on pedigrees that went back to the original settlers, before he died in April 1943.

The information he collected was never put together into one study. Instead, pieces of it were used in other studies. In 1954, Dr. C. Nash Herndon - Allan's protege - wrote a paper, "Intelligence in Family Groups in the Blue Ridge Mountains," that was based partly on Allan's study. The study used the IQ data that Allan had collected from 223 people in 86 families during the Watauga survey to study IQ differences between husbands and wives.

Herndon also used the pedigrees and Allan's other work to study the effects of marriages between cousins. The rate of first-cousin marriages in Watauga County was about 6.72 percent, well above the American and European averages of .2 to 1.03 percent. The increased rate did not affect the intelligence of the population, the study found.

Doing the studies took a special kind of person, Allan told Frederick Osborn, a nationally known eugenicist, in a letter dated Jan. 24, 1939.

"If you and Professor Dunn picked a man, with your gift of friendliness, it would be smooth sailing but a man with ... shyness and inhibitions, for instance, would be mistaken for one of General Sherman's stragglers and strung up. At this stage someone who has been trained to deal with the plain, common Scotch-Irish farmer would be more useful than someone trained in science."

Allan spoke from experience. While he generally got along well with the people he went to study, he had a few interesting moments, as Herndon and Allan's daughter recalled during an oral-history interview in 1976 arranged by the archives.

He was once mistaken for a Nazi spy. He had dogs set on him at least once. Another time, a dog helped him gain entry to a house. When the family dog didn't bite Allan while he waited at the gate for 10 minutes, the woman of the house agreed to talk to him.

But some Watauga County residents were suspicious of Allan's work for other reasons than simply being wary of strangers, especially when he attempted to take saliva samples with a small strip of paper.

"Well, I understand that word got around in the back country that they were sneaking around and sterilizing them with these things and the people didn't want to take it. At least some of them didn't," said Allan's daughter, Elizabeth Allan Berger, in the oral-history interviews.

"They thought that this was going to keep them from having children and ... so the men would, were, taking off but the ladies were crowding around," Herndon said.

Advocate: Wake Forest president embraced eugenics movement

By John Railey
JOURNAL REPORTER

William Louis Poteat fought for causes ranging from child-labor reform to humane care of the mentally ill. He earned a progressive reputation that blossomed nationwide from his post as the president of Wake Forest College.

Poteat died in 1938, years before the college moved to Winston-Salem and grew into a university, but his name lives on at the school. Wake historians still talk about the legacy of "Billy" Poteat, including how he made a strong case for academic freedom by openly teaching evolution in his biology classes in the 1920s.

Another position taken by Poteat is not discussed as much. He was an early and vocal supporter of the eugenics movement. He embraced it long before others and stood by it long after many others backed off. He gave his students pamphlets advocating sterilization for human betterment and in speeches across the South he pushed the idea of sterilizing the unfit.

"(T)here can be no doubt that we are ready for the application of negative eugenics, that is, restrictive mating for the elimination of the obviously unfit," he told a group of Baptist educators in 1921.

Poteat was a devout Baptist born shortly before the start of the Civil War. He worked passionately through the first third of the 20th century, exploring a cavalcade of new ideas - including eugenics. The seemingly contradictory sides of Poteat's life raise an old and thorny question for historians. Should great men be seen in the context of their times, or should more be expected of them?

"It's what some people call 'the present-tense myth,' that we are trying to make him see what we now see," said McLeod Bryan of Winston-Salem, 82, a retired religion professor at Wake Forest University.

Bryan, who did his master's thesis on Poteat while at Wake Forest College, said that Poteat was a paragon of another myth - the scientific myth "that science is going to cure everything. And he was a perfect example of that, he swallowed it hook, line and sinker," Bryan said.

For Poteat, eugenics "would have been the latest or most scientific approach to making society better," said Randal Hall, the author of William Louis Poteat: A Leader of the Progressive-Era South, published in 2000.

Poteat, a Wake Forest graduate, embraced eugenics as he took leadership of the school during the early 1900s. He continued to teach as president.

The Eugenics Board of North Carolina, the state's vehicle for ordering sterilizations, wasn't formed until 1933. Poteat didn't participate in its creation; that was done by an act of the legislature. But as an influential leader revered across the state, he helped lay the groundwork for its acceptance.

At a YMCA retreat in Blue Ridge, N.C., Hall writes, Poteat taught classes on eugenics. But, as Hall notes, "most of his work on behalf of eugenics took the form of speeches vainly advocating the idea."

The speech Poteat gave to the Southern Baptist Education Association in 1921 was typical. "The feebleminded, the insane, the epileptic, the inebriate, the congenital defective of any type, and the victim of chronic contagious diseases ought to be denied the opportunity of perpetuating their kind to the inevitable deterioration of the race," he said.

Both Hall and Bryan said that Poteat's support of eugenics and progressive causes was not an unusual combination for his time. "For a time, I think, the (eugenics) position had the support of many but not all progressive voices," said Hall, an admissions officer and a history professor at Wake Forest. "Not just in the South, but in the rest of the country."

Hall said he was "a little disappointed" when he came across Poteat's connection to eugenics while doing research for his book. "Where he doesn't change is the late '20s and early '30s. By then, there were voices in opposition to eugenics that he just doesn't acknowledge."

Neither Hall nor Bryan thinks that Poteat's stance on eugenics diminishes his reputation. "I think we have to acknowledge that even our heroes are human and voices for their time," Hall said.

Castration: Files suggest that punishment was often the aim

By Kevin Begos
JOURNAL REPORTER

The Eugenics Board of North Carolina acknowledged three diagnoses - epilepsy, mental illness and feeblemindedness - when approving sterilization petitions. But did it also approve sterilizations just as experiments?

The *Winston-Salem Journal* reviewed thousands of records of the eugenics board, but more are still under seal in the N.C. State Archives, and hints about experimental sterilization appear in other places.

"The operation for sterilization by castration was performed for Robert, January 5, 1934. Though the operation was entirely within the law, it was performed in a somewhat experimental manner.

"It was sort of an experiment in this way: that a follow-up study of Robert was requested to see what effects the operation might have upon his actions, so they might go ahead and use the same operation for other cases which were similar or closely related."

Those passages are from a 1936 thesis written by J. McLean Benson for the University of North Carolina at Chapel Hill. Benson was given access to a selection of early eugenics board records and also to records from the Orange County Department of Public Welfare.

Robert, a black man, had been convicted of indecent exposure, Benson wrote, and since sterilization "would not stop" his actions, the eugenics board ordered him castrated. The records of Robert's case from the eugenics board have not been made public.

There were 65 castrations performed under orders from the eugenics board, and the records suggest that such operations were meant as punishment in addition to the goal of preventing some people from reproducing. The last documented castration was done between July 1956 and June 1957.

A suggestion of indecent sexual behavior by a black man was more likely to result in castration, and in 1937 the eugenics board ordered an asexualization operation (hysterectomy) for a 38-year-old white woman who had just had a mixed-race child.

Another case with fragmentary records is of a 10-year-old boy who was castrated at the State Hospital in Goldsboro in 1935. The reason for the operation was given as "Low grade imbecile," without any explanation of why castration was necessary, instead of a vasectomy. The boy died of tuberculosis less than a year later.

Selling a Solution
Group founded by Hanes, others sent sterilization in new direction

By Kevin Begos
JOURNAL REPORTER

James G. Hanes was a master at selling hosiery, and he was just as successful at selling an expanded program of eugenic sterilization to the people of North Carolina.

In the spring of 1947, Hanes helped found the Human Betterment League of North Carolina, bringing cash, a Manhattan advertising agency and slick mass mailers to promote an idea that almost every other state was writing off as a mistake.

North Carolina had been doing sterilizations since 1929, but in a haphazard way. By 1947 the legislature still hadn't authorized money for a full-time clerk or even a permanent office for the Eugenics Board of North Carolina. The flawed science of eugenics made exaggerated claims that mental illness, genetic defects and social ills could be eliminated by sterilization.

As in many states, officials had pushed a eugenics program in the 1920s and 1930s, but the idea had lost support on both the political and scientific fronts. Sterilizations in North Carolina had peaked at 202 in 1938 and then fallen to 117 in 1945.

The Human Betterment League changed all that. It gave the eugenics board new legitimacy and political clout at a time when it needed it most.

And clout was something that James G. Hanes and the Winston-Salem elite had in abundance. Hanes Hosiery was the largest-selling nationally advertised brand, and R.J. Reynolds Tobacco Co. was the biggest tobacco company in the nation. Hanes also served on the Forsyth

County Board of Commissioners for 22 years, 20 of those as its chairman.

"I think Winston-Salem (of that era) was a little bit exceptional, in North Carolina or in the South, really," said Bob Korstad, a professor of history at Duke University. His book about race and labor relations in Winston-Salem during the 1940s is scheduled to be published next year.

"There is this small intertwined elite that sees Winston-Salem as kind of a feudal manor in a way. These people had an incredible amount of wealth, and a lot of political influence — even on the national level. They just assumed that they were the best men, and that they knew what was right for people."

Good intentions

Although there was some individual variation of opinion as to the value of the distribution of material containing possible controversial statements, there was unanimous agreement that valid educational publicity is essential.
— Human Betterment League minutes, 1947

Like any well-organized business, the Human Betterment League seemed to have every base covered. Hanes was joined by Alice Shelton Gray, a trained nurse and another member of the local elite. Further bolstering the ranks were Dr. C. Nash Herndon, a leader in the medical-genetics department at the Bowman Gray School of Medicine; Dr. Clarence Gamble, the Harvard-educated heir to the Procter & Gamble fortune and an important figure in the fledgling birth control movement; and Dr. A.M. Jordan, a professor of psychology at the University of North Carolina at Chapel Hill. The group had money, political connections and a conviction it was doing the right thing.

The founding members are dead now, but records from the era describe two key events in the evolution of the Human Betterment League. During World War II, a large number of draftees from North Carolina had been rejected by the military as "mentally unfit." Hanes had been upset by the news and decided to identify the problem and find a solution.

Perhaps unknown to Hanes, the "mentally unfit" label was flawed. It was true that many men from the state had been rejected by the military, but that catch-all category included legions of farm boys — black and white — who had strong backs and common sense but so little schooling that they couldn't read or write. In response, the Army designed a new IQ test that used pictures instead of words to deal with a problem that was hardly unique to North Carolina.

Still, Hanes charged ahead, paying for a massive survey that gave IQ tests to about 10,000 students in the Winston-Salem school system. The results suggested that alarming numbers of the children were "feebleminded," and soon there were editorials appearing in the *Winston-Salem Journal* suggesting that there was a problem across the state.

"Miss Wulkop served tea"

Gamble had been promoting birth control since the late 1930s, and he met Alice Gray through those campaigns. Gamble was sure that eugenic sterilization was a good idea, but after World War II few states were willing to consider the kind of aggressive program that he wanted. Gamble contributed time, money and a keen public-relations sense to the Human Betterment League. He also paid for most of the sterilizations in Orange County during one year, and he paid for the research that went into the book Sterilization in North Carolina, written by researcher Moya Woodside.

A mailing list of circ. 40,000 has been prepared, consisting of upper class college students, faculty members, physicians, nurses, ministers, public officials, etc. To them have been sent 110,000 items.
— Human Betterment League minutes, 1948

"They were willing to start this huge publicity campaign about why North Carolina should continue the program, and why sterilization was so incredibly valuable even though all of science by that time started saying

"No, this is really the wrong thing to do, and this is really unscientific,'" said Johanna Schoen, an assistant professor of women's history at the University of Iowa who has been researching the eugenics board for more than 10 years.

Though he considered himself a progressive believer in birth control, Gamble had an especially nasty edge, Schoen said.

He composed "poems" that extolled sterilization for "morons." In one, a young woman and man are led toward sterilization, and Gamble concludes that "the North Carolina MORONS lived happily ever after."

"It's a very paternalistic model," Schoen said. "(Women) certainly weren't supposed to choose when to use birth control or when not to use birth control, or when to be sterilized or when not to be sterilized. The model was, the physician knows best."

There are signs that some in North Carolina were horrified by Gamble. He suggested that the N.C. Mental Hygiene Society use his poem to promote the sterilization campaign; when its leaders firmly declined, Gamble was puzzled.

"Your unfavorable criticism of the story of the two moron families interested me," Gamble wrote in a January 1947 letter. "It will be helpful if you can tell me the reasons behind this."

But at least in the beginning, the Winston-Salem elite accepted Gamble into the fold, and genteel manners may have helped smooth any rough edges.

Dr. Gamble spoke informally, reviewing the survey of students in Winston-Salem schools made by psychiatrists working under the supervision of Dr. A.M. Jordan, financed by Mr. James G. Hanes. Dr. Jordan's findings are arresting and intensely interesting. As presented by Dr. Gamble, with a wealth of collateral comment, they were received with keen concern and an animated discussion followed. At the adjournment, Miss Wulkop served tea to those present.
— Human Betterment League minutes, 1948

The motivations of the founding members of the Human Betterment League are open to speculation, but their success in getting the attention of others around the state is clear.

Dr. Ellen Winston, the head of the state Department of Social Welfare, noted "the special interest of Mr. Hanes in regard to a more intensive sterilization program" at a 1948 meeting of the eugenics board. The same year, George Lawrence, a member of the Human Betterment League, noted "a greatly increased interest among Welfare Workers in the use of sterilization for their clients and an even more significant change in attitude among surgeons and public health doctors."

Turkey, politics and Mr. Hanes

Redge Hanes, a grandson of James G. Hanes, remembers the regular Sunday lunch with turkey at his grandparents' house.

"I recall he asked once what we studied in school that week, and I told him Mrs. Barber said Franklin D. Roosevelt was the greatest American president of the 20th century." Hanes said his grandfather "pounded his hand down and almost broke the table — he said that was a lie."

At least once, the conversation turned toward the Human Betterment League.

"The idea was one of planning and stability. There was no (birth-control) counseling, no guidance then. He was concerned at the numbers of children in poor families," Hanes said recently. "There wasn't any mandatory anything. It was that you do have an option.

"What he wanted to do was to give them an option to stop having children, because he saw that as a terrible drain on the public coffers," Hanes said.

Korstad said that the elite of the city "really saw themselves as progressive social engineers."

"They're able to exercise power in a lot (of) subtle ways," he said. "They were obviously pretty sophisticated in the way they did it. They don't resort to the kind of raw power and violence that people did in other parts of the South."

A huge slum-clearance project was in progress in the city during the late 1940s, Korstad said, and the push for an expanded sterilization program might have been a response to changes in society.

"Think about it also in the context of how the postwar labor needs changed pretty dramatically. You've got this tremendous mechanization (in industry) which is creating, from their point of view, a surplus black population. They don't need sharecroppers families with ten kids anymore," Korstad said.

A powerful elite did decide what to do and when to do it, but that wasn't all bad, said the Rev. Jerry Drayton, who has been active in the civil-rights movement in Winston-Salem for more than 50 years.

"The circumstances of the time were such that that power was needed to get things done," said Drayton, who added that with regard to racial issues James G. Hanes used his clout to "do a great job."

"He started the Urban League. He personally paid for the executive director and he formed a board of directors — 10 blacks and 10 whites," Drayton said. "He was active with us, the NAACP, quietly. He made a great contribution toward the opening up of Winston-Salem."

Out of their hands

Mr. Lawrence recently interviewed Mr. J. B. Moore, the newly appointed Director of State Prisons, and found him keenly interested in better enforcement of the sterilization law, especially among women prisoners.
— Human Betterment League minutes, 1949

Within a year of its founding, the Human Betterment League seemed to play a major role in changing the course of the North Carolina sterilization program. Sterilization numbers started to increase dramatically, and by the early 1950s the state had the highest per capita rate in the country.

The minutes of the group note that Hanes gave a luncheon to "some 30 prominent men of Winston-Salem," Alice Shelton Gray gave a tea for a large group of Winston-Salem women and Gamble got help from Madison Avenue.

Dr. Gamble ... enclosed a letter of solicitation he has had prepared by James Gray, Inc., professional ad writers in New York.... He congratulated us upon the showing of increased sterilization in N.C.
— Human Betterment League minutes, 1951

Gamble decreased his involvement with the Human Betterment League and North Carolina by the mid-1950s, and he turned his attention to international birth-control and sterilization programs.

"It becomes this self-propelled thing," Schoen said. "There are enough people in North Carolina who are committed to it that they don't need a Gamble anymore. The legislature is happy to keep funding (the eugenics board)."

By 1957, the Human Betterment League had sent out more than 575,000 pieces of mail promoting the sterilization program. In the coming years, increasing numbers of sterilizations would be done on people from the general population, especially blacks and women.

Psychiatrists started to openly question IQ tests and the blanket use of the term "feebleminded" to describe a variety of social and medical problems, but outside of the tightly controlled environment of Winston-Salem, others were pushing the program to extremes that worried even some members of the Human Betterment League.

Sterilization as a social punishment Dr. Herndon believes is the wrong approach, social ills should not be confused with mental ills. It is most important to separate illiteracy from mental retardation in the handling of this problem.
—Human Betterment League minutes, 1959

Herndon, of the Bowman Gray School of Medicine, was talking about a bill in the legislature that would have required sterilization for mothers with more than two illegitimate children. Herndon suggested that the Human Betterment League issue a statement distancing itself from the legislation, but "after some discussion it was agreed not to attempt to do this," the minutes of the meeting read.

The bill never passed, but the eugenics board promoted the same idea. Welfare rolls were rising throughout the state, and sterilization was seen as one way of reducing costs.

Though the North Carolina sterilization program became racially imbalanced by the late 1950s, Redge Hanes said that outcome was never part of the agenda of the Human Betterment League or consistent with his grandfather's progressive views on race.

"If the discussion (now) is that abuses took place, abuses take place in every program," Redge Hanes said, adding that he had "not the slightest idea" if his grandfather knew of those problems.

Dr. Charles Hendricks, who came to Winston-Salem in 1968, joined the Human Betterment League a year or two later and served as president. Hendricks, who now lives in Chapel Hill, said that there was no talk of sterilization at league meetings in the later years, but that it was clear that James G. Hanes had been the moving spirit of the group.

Hendricks described Hanes as a "good, generous man," but doubts that he understood how complex the issues of sterilization and genetics were. "I think he (Hanes) thought he would solve the problems by getting a few people together, and they'll do the sterilizations, and it will help society," said Hendricks. "Obviously, in retrospect, that was a desperately shortsighted view."

Whatever the intentions, Schoen said, "Gamble and the Human Betterment League helped something happen in North Carolina that was happening almost nowhere else."

The sterilization program may have saved taxpayers money, she said, but in the end it failed as social policy.

"The (N.C.) legislature might be willing to fund eugenic sterilization, but the legislature was not yet willing to fund many other things ... in terms of addressing those problems that eugenic sterilization is supposed to address," Schoen said. "Incredible poverty. Lack of education. It's like this Band-Aid solution — sterilization is the easiest way to deal with it."

Redge Hanes said his grandfather may have realized that, too. At another Sunday lunch discussion in the late 1950s, the Human Betterment League came up again.

"To a certain extent it was doomed from the start. That was the kind of discussion I recall," Redge Hanes said of comments his grandfather and father made. "It was that it (the league) just didn't work — the point that my grandfather was interested in was simply not achievable."

But perhaps because Hanes and others had promoted the sterilization program so successfully, the attitudes that went with it were hard to stop.

When the Human Betterment League held a meeting at the Hotel Robert E. Lee in 1964, Dr. Henry O'Roark, the director of the Forsyth Family Planning Clinic, noted "that from $1,500 to $2,000 can be saved by the county for the prevention of birth of each illegitimate child."

The eugenics board continued to order sterilizations until 1974. Toward the end of the program 99 percent of the operations were on women and more than 60 percent of those on black women. Korstad said that the old culture of power in Winston-Salem cannot be described as simply one where good intentions sometimes went wrong. It "wasn't an environment where you could have a disagreement with these guys and what they were up to," he said.

"Winston-Salem as a community, they need to confront some of this," Korstad said.

By the early 1970s the Human Betterment League shifted its focus to producing educational materials on birth control and genetic counseling. In 1984 it became the Human Genetics League. It finally ceased all work in 1988.

'It Ain't Fair'

Old IQ score helped social workers get reluctant teen-ager sterilized

By John Railey
JOURNAL REPORTER

WANCHESE– Bertha Dale Midgett Hymes comes from a place where generations of locals have spilled blood and sweat just surviving.

Hymes, 52, grew up in Wanchese, a fishing village on Roanoke Island where folks take pride in hard work. It is also a place where people struggle to get by and government handouts are sometimes grudgingly accepted. Hymes' family accepted that support. When the social workers who administered the support learned that Hymes, 16, was pregnant and unmarried, they pushed to have the state sterilize her after the birth of her child, saying that she was mentally retarded.

Hymes, who has a heavy speech impediment, doesn't usually talk much about what happened to her. "It's embarrassing, you know. People don't want to hear it. Everybody calls me names, makes fun of me, how I talk," she said.

But with the details of what happened to her and thousands of other North Carolina girls and women emerging, Hymes now wants to tell her story and strains to find the words. "What God gives to you, they take away from you.... I don't think it's right. It ain't fair. I think the government should apologize to me and everyone else."

Hymes is white; the majority of the others ordered to undergo sterilization by the Eugenics Board of North Carolina in the 1960s were black. But like many of those sterilized, she was young and poor and didn't have much chance to voice her objections.

Some social workers - including Hymes' current case worker - now dismiss the work of that board. "I think it was based on junk science," said Sue Judge of the Dare County Department of Social Services. "Basically, it never worked and it certainly worked hardships on individuals."

Hymes bristles as she remembers the operation she underwent at a hospital a half-hour's drive inland from her home. "They cut me. They tied my tubes so I couldn't have any more babies," she said.

Hymes' daughter, Frances Scarborough, 35, sympathizes with her mother. "I don't see where they had a right to take that option away from her," said Scarborough, who is of normal intelligence.

A recommendation of sterilization

A short woman of medium build with grayish-brown hair, Hymes lives in East Lake, about 20 miles from Wanchese. Both places are just minutes away from the condo-studded beaches of Nags Head but worlds away in lifestyle. Most of the folks from East Lake and Wanchese eke out a living working on the same water and sand that the tourists play on.

Hymes cooked in fast-food restaurants until taking disability a few years ago. She said she is not mentally retarded, though she has received disability checks from the government for being mentally handicapped. She is illiterate, she said, but can write her name.

She is one of eight children of a ferryboat captain and a homemaker, Dorothy Midgett. Her parents divorced when she was young.

"It was a hard situation when we were growing up," said Hymes' sister, Elsie Sanderling, 57. "We were really like a bunch of animals."

Hymes said that a married man got her pregnant when she was 16. "He said he ain't got me that way. I said, 'I didn't get that way by myself.'... They kicked me out of school because I got pregnant."

Records from the Dare County Department of Public Welfare and from the eugenics board tell the rest of the story, one of Hymes' mother being pushed to consent to the sterilization.

Worker told her that services were available and it was very apparent that Mrs. Midgett was quite hostile toward worker and went on to say that she did not want the Welfare Department to bother anything about it and they would not be willing to accept any services from the agency ... She became more hostile and said that she wanted the Welfare to stay out of her affairs, that she did not want a thing from us, and that she did not invite worker to come back.... When it became apparent that worker could not talk further with Mrs. Midgett, worker left.

– Case notes of Doris Bonner, a social worker for the
Dare County Department of Public Welfare, July 1967

Through the summer and fall of 1967, Bonner visited the Midgett home, encouraging Hymes' mother to let Dale be sterilized. "I just didn't like the idea of it," Dorothy Midgett, 82, said recently.

According to Bonner's notes, the visits started after Hymes' family doctor, W.W. Harvey Jr. of Manteo, called the welfare department about her case. Harvey was concerned because Hymes was pregnant and mentally retarded, Bonner wrote. Harvey was hopeful that she could have an abortion but then realized she was 3 1/2 months pregnant - too far along to undergo that procedure.

Harvey said that Hymes should be urged to give the baby up for adoption, and that she should be sterilized after giving birth, Bonner wrote.

Bonner and her boss, department director Goldie Meekins, began preparing a sterilization petition to submit to the eugenics board. As part of that petition, they needed a current IQ score.

Bonner and Meekins had the results from an IQ test that Hymes took when she was 11. The test determined that she had a mental age of 5 and that her IQ was 48, well below the cutoff of 70 used to classify the mentally retarded.

A potential problem had surfaced, though. A few years earlier, a lawyer for the eugenics board expressed reservations about using old IQ test results on petitions because of the threat of lawsuits.

But in a letter to Bonner and Meekins, the executive secretary of the eugenics board, Sue Casebolt, said that the old test was all that was needed.

The eugenics board still needed Hymes' mother to sign a consent form. As a minor, Hymes would have no say in the matter.

'Favorable action'

(Hymes) is rather pathetic looking as she is now gaining considerable weight and does not seem to feel too well. However, she seemed rather cheerful. She was dressed in a not-too-clean dark green cotton dress which was full. She showed worker a new brown cotton maternity dress which her sister-in-law had made for her. She was quite thrilled with the new dress, and it seemed more pathetic that she does not really realize her condition and what can happen in the future to her and the baby to be born – Bonner case notes, Oct. 4, 1967

Hymes' mother seemed willing to have the sterilization done, Bonner wrote. Midgett said that wasn't true and once even threatened to "call the law" on Bonner when she visited.

Midgett finally agreed to sign the consent form just as Hymes entered Columbia Memorial Hospital to deliver her baby. Midgett said she did so because she was told that her daughter was fragile and having another baby could threaten her life. But medical records do not say that, nor does an exchange of letters between Bonner and Meekins in Manteo and Casebolt in Raleigh that suggests a rush at getting the sterilization done.

The exchange ended with a short note from Casebolt.

In special meeting, the Eugenics Board took favorable action today on the petition for sterilization of Bertha Dale Midgett.

On Dec. 2 at 11:14 p.m., Hymes gave birth to Frances. She weighed 5 pounds, 15 ounces. The delivery doctor, Robert Albanese Sr., wrote in hospital notes that Hymes underwent another operation three days later.

This patient had been cleared for a sterilization procedure by the North Carolina Eugenics Board. Consequently, on the 5th of December, she was taken to the operating room where under a general anesthesia a bilateral partial salpingectomy and tubal ligation were performed.

Dr. Arthur Bradsher, a surgeon from nearby Windsor, did the operation. Hymes said she talked to Bradsher beforehand. "He said, 'I'm going to sterilize you. Your mama has signed the papers.'

"I said, 'I don't want to be sterilized.' He said, 'You can't do nothing about it.'"

Like many surgeons who performed sterilizations for the eugenics board, Bradsher has since died.

"Bradsher, in fairness to him, he was just trying to help us out," Albanese said recently. "If you interviewed 100 people in those days, I don't think you could have found one who thought it (sterilization) was a bad idea."

Albanese said that although he didn't do sterilizations - he wasn't a surgeon - he understood the reasoning behind them. "There was just unbelievable poverty there; you just couldn't believe it. The idea of them bringing more children into the world in a situation like that just didn't make sense."

In her notes, Bonner wrote that she talked to Albanese about the operation and that he was hopeful that it could be done. Albanese, who now lives in Martinsville, Va., said he couldn't remember Hymes' case but "I have no doubt that I went along with that."

"Looking back ... sure I would regret it," he said. In the scheme of things, he said, sterilizations were a good idea "but if you consider individual rights, it was not a good idea."

'No justice'

In the days after the sterilization, Hymes "complained big-time," Sanderling said. "She cried. I thought at first Dale was going to have a nervous breakdown."

Hymes' mother raised Frances. Hymes lived with them in Wanchese, and Frances at first thought that her mother was an older sister, family members said.

Hymes got married and moved out of the house. Her first marriage ended in divorce, she said. Hymes said she drank for a while, then quit.

"I got married again," Hymes said. "He wanted kids. I told him I couldn't have no more kids because my tubes were tied."

A widow, Hymes subsists on payments from her husband's Social Security account. She owns a mobile home in East Lake, where she lives with her daughter and her two granddaughters.

Sometimes her thoughts turn to her sterilization and the circumstances that led up to it. She is all but alone in those memories.

Most of those involved in having Hymes sterilized are dead now - including Meekins and Bonner. Judge, Hymes' current social worker, joined the county welfare department after the legislature disbanded the eugenics board in 1974. For a while she worked under Bonner, who became the head of the department.

Bonner "certainly tried to do her best for folks," Judge said. Bonner didn't go "out searching for people" to sterilize, she said.

"I think that Mrs. Bonner responded to family requests when families would come in and say to her, 'I have a daughter who's mentally retarded. She's pregnant. I don't want this.'"

Judge declined to comment further.

Hymes looks at her granddaughters, and she thinks of what happened to her and other girls when they weren't much older than them. "It's terrible what they did to the kids. It's hush-hush," she said.

"They didn't give me no justice."

City's kids put to the test in '48

UNC professor's work, underwritten by James G. Hanes, may have been fodder for sterilization campaigning

By Danielle Deaver
JOURNAL REPORTER

When members of the Eugenics Board of North Carolina were trying to decide whom to sterilize, they almost always relied on IQ scores as the one objective measure they could count on among a number of other considerations.

But even 50 years ago, researchers had doubts about what intelligence-quotient tests - including the Stanford-Binet IQ test used by the eugenics board – could accurately tell them. Scientists knew that environment, nutrition, early-childhood enrichment and even cultural bias on the test played a role in determining IQ.

The test has undergone radical changes since its inception in the early 1900s. Despite the doubts, people - including members of the eugenics board - still continued to see the test as something that could reveal all about a person's potential.

IQ tests can't meet that expectation, said Bruce Bracken, the president of the International Testing Commission and a professor of education at the College of William & Mary in Williamsburg, Va.

"We have no measures of potential.... The tests we use merely measure a person's present capacity. They may have previously worked at a different level," he said. "The best use of intelligence tests is just to assess what the person's current ability is."

A difference in IQ scores between the races was noted in a 1948 study of Winston-Salem schoolchildren conducted by A.M. Jordan, a professor at the University of North Carolina at Chapel Hill.

Using money from Winston-Salem textile magnate James G. Hanes, Jordan tested the IQs of 95 percent of the elementary-school students - about 10,000 children – in the city. Jordan's team decided to use an IQ score of 60 as the dividing line between retarded and normal instead of the standard 70 used in other studies.

Even with the lower dividing line, the results were not "entire successful," Jordan wrote, because researchers found that the children tested in Winston-Salem had lower average IQs than their counterparts in other parts of the country. Black children tested in the city had lower scores than white students.

Individual tests confirmed that about 60 percent of the children who tested below 70 actually had IQs that low, Jordan wrote. The rest of the children had been scored too low.

In addition to compiling information about the children of Winston-Salem, Jordan was also evaluating the use of group tests to determine which children needed more help.

But he may have been doing something more. Hanes and Dr. Clarence Gamble, a wealthy supporter of the eugenics movement, may have also been trying to use the tests to promote further sterilizations, according to a thesis by William Van Essendelft, a graduate student at the University of Minnesota who was studying a group called the Voluntary Sterilization League.

"In Winston-Salem and in Orange County, North Carolina, the field committee had participated in testing projects to identify school age children who should be considered for sterilization," Van Essendelft wrote.

Doctors with the Bowman Gray School of Medicine may have been involved in the testing and may have tried to use the results to increase public interest in sterilization, Van Essendelft wrote.

"The medical school has a long history of interest in eugenics and had compiled extensive histories of families carrying inheritable disease. In 1946, Dr. C. Nash Herndon ... made a statement to the press on the use of sterilization to prevent the spread of inheritable diseases. That release coincided with the testing program and was part of an effort to use the tests to stimulate interest in sterilization," Van Essendelft wrote.

In the end, seven children were identified as candidates for sterilization, a number that disappointed Hanes and Gamble. It is not clear whether the seven children were ever sterilized.

The group apparently didn't use typical eugenic board standards - which made anyone with an IQ under 70 a candidate for sterilization - to decide who should have the operation. If the results of Jordan's study were the only factors used to determine who should be sterilized among the 10,000 children tested in 1948, 218 white children and 526 black children would have been sterilized.

Jordan saw the wide discrepancy between black and white children as a sign of a problem with the test.

"One gains immediately the impression that the test was not fair to them or that the fact that the tester was white could have made some difference," he wrote.

Jordan blamed the low performance on the fact that most of the black children's parents had less education than white parents, and that there was a limit on the expectations people had for black children - two theories that are still considered valid today by experts in the field.

But other modern researchers think that the difference is caused by a number of other factors, not the least of which is how the test is constructed.

The IQ test was made up by whites and tested on white children until an update in 1972. So the version that the eugenics board used was based on the experiences and abilities of white administrators and children.

This could have been part of the problem, said Frank Wood, a professor of neuropsychology at Wake Forest University Baptist Medical Center. IQ tests have always been criticized for posing questions that the majority culture would have an easier time answering.

"Culture-fair items are supposed to be items that are just as regularly solved by people in one culture as another. People who grew up in a vegetarian culture might not have learned very much about cooking meat, for example," Wood said.

Intelligence-quotient tests have been around since the early 1900s. They were developed by French doctors to separate developmentally disabled children from other children who were entering school.

The Stanford-Binet is a one-on-one test that has to be administered by a trained professional. It takes more than an hour to give, and the scoring takes longer yet.

The test was created in 1905, and modified in 1916 when it was given the name Stanford-Binet.

There have been four revisions done since then, with a fifth scheduled to come out in 2003.

The Stanford-Binet has a series of questions in different subject areas. Some deal with reason and comprehension. A typical question asks children to name the similarities and differences between an orange and baseball.

Others test general knowledge by asking such questions as naming the days of the week.

Some test spatial intelligence by looking at a child's ability to draw a picture of an object or a design placed in front of him or her.

"When you add it up across these different ways of testing it you get to feel that you are getting a more general sense than you would any other way," Wood said.

Even if IQ research has led to some insight during the past 100 years, several questions still remain, such as what creates intelligence in a person.

"People tried to look at it to see if it was inherited. That turned out to be only minorly interesting. Yes, it runs in families, but familial connections, including genetics, have never been the whole of IQ," Wood said.

Most researchers agree that such factors as early-childhood enrichment and nutrition play a role in the formulation of IQ. But they disagree on many other points.

City's kids put to the test

During the past 100 years, IQ scores have increased by about 15 points. The average score is still 100, but every time the average is reset, a new scale also has to be recalculated to take the difference into account.

Though test experts cannot pinpoint a reason for the increase, some point to better schooling and the effects of living in a more complex society as possible explanations.

"Our overall ability to do that as a population has increased slowly but steadily over the past 100 years. That's one reason you have to restandardize the test every few years. If you've got an IQ of a certain level on a test you took 20 years ago, that same level of skill would get you a lower score now," Wood said.

One of the most controversial aspects of the test has always been the difference in scores between black and white children.

Blacks score an average of 15 points below whites on the tests. That difference was the subject of a 1994 book, The Bell Curve, which suggested that the discrepancy was due to genetic differences between the races.

People still argue over the book's ideas. Bracken thinks that the reason for the difference is more complex than a simple genetic difference.

"If you look at the nature-nurture issue, some people believe that the differences are genetically related, some believe the differences are almost purely environmental. Most people agree there is a combination between nature and nurture and we also have the effects of cumulative deficiency. When you compound socioeconomic deficit over generations, you get a compounding effect," Bracken said.

No matter how they were constructed, IQ tests should not be used to decide who could have children, Bracken said.

"I think that's despicable. There are many good reasons to use IQ tests and there are obviously extensions that are inappropriate," he said. "Deciding who should reproduce should not be one of those uses."

ALL ABOARD:
Newspapers jumped on sterilization bandwagon

By Kevin Begos
JOURNAL REPORTER

North Carolina's newspapers played a key role in supporting the state's eugenic sterilization program, providing a barrage of flattering editorials, features and news stories that helped cut short serious debate on the issue.

"The danger is in the moron group which includes a host of physically attractive individuals whose IQs are lower than a January thermometer reading. Among other things, they breed like mink," wrote Chester Davis in a full-page feature for the combined Sunday edition of the *Winston-Salem Journal and Sentinel* in 1948 under the headline "The Case for Sterilization - Quality Versus Quantity."

The *Journal and Sentinel* ran dozens of articles in support of the sterilization program, and that was wrong, said Jon H. Witherspoon, the current publisher of the Journal.

"Certainly, any role that the *Journal* played in advocating forced sterilization is reprehensible," Witherspoon said. "Taken in the context of the era, such sentiments were probably more 'middle of the road,' but nonetheless the sterilizations violated the human and civil rights of many people, over 7,000 in North Carolina."

Instead of asking hard questions about the sterilization program, the Journal ran fluff pieces that led readers to believe that all was well within the flawed program.

"I regret that the Journal, in its past, played a role in legitimizing these barbaric activities. On behalf of the *Journal*, I apologize for the paper's part in depriving these individuals of their basic human rights," Witherspoon said.

The *Journal and Sentinel* promoted the sterilization program, but it wasn't the first or the only paper to do so.

As millions of Americans struggled to survive in the depths of the Depression, an editorial from the *Raleigh News & Observer* suggested that the most pressing problems of the day were eugenic rather than economic.

"We cannot make a better world if we deliberately give our substance to subsidizing the production of the least worthy stock among men," read part of the 1935 editorial, which the Eugenics Board of North Carolina reproduced in its biennial report. Lamenting that the men "who do the world's work and pay the world's taxes" were having smaller families, and that "the lowest orders of humanity" were increasing, the editorial suggested that solving that problem "is far more important than the solving of any of the immediate problems of economic depression."

"People viewed this as progressive, as science," said David Goldfield, a professor of history at the University of North Carolina at Charlotte, and overlooked the fact that scientists can make mistakes, too.

"The '30s and '40s were decades of intense conflict between modernism and tradition in the South, with religious fundamentalism battling Darwin, and Howard W. Odum and his Regionalists at Chapel Hill trying to study the South's dark closets of lynching and poverty," Goldfield said. "The media, typically on the side of the modernists, hopped on the science bandwagon."

Davis' prose about sterilization for the Journal and Sentinel was no accident, according to Bob Korstad, a professor of history at Duke University who has a forthcoming book on race relations in Winston-Salem during the late 1940s. The Journal and Sentinel embraced the cause of eugenic sterilization with fervor. That support was clearly tied to the fact that the elite of the city had founded the pro-sterilization Human

Betterment League of North Carolina in 1947, Korstad said.

Gordon Gray owned the Journal and the Sentinel at the time, and his cousin, Alice Shelton Gray, was a founding member of the Human Betterment League.

Davis was a graduate of Georgetown University and Harvard Law School, and served as a special agent with the FBI from 1940 to 1946. A graceful writer and a talented reporter who won national awards, he did Sunday investigative reports on many key issues, from school reform to the environment.

The culture of the time meant that Davis' features were virtually certain to reflect the views of Winston-Salem's elite, Korstad said.

"(Davis did) it in a somewhat sophisticated and liberal way. But it was all about controlling the discussion," he said.

The closeness between what the elite was interested in and what the paper supported editorially was sometimes blatantly obvious. The Human Betterment League was incorporated March 22, 1947 - and the very next day, there was a favorable mention about sterilization on the editorial page of the Sunday *Journal and Sentinel*.

In 1948 and 1949, the Sunday *Journal and Sentinel* ran dozens of guest columns under the headline "It Could Happen Here." Written by Elsie Wulkop, the secretary of the Human Betterment League, they were filled with sentimental stories about the handicapped, the poor and the feebleminded.

The propaganda wasn't confined to the editorial pages. Papers in the state also ran articles extolling the sterilization program, and the Eugenics Board took note of the help, mentioning a 1950 article that ran in the *Asheville Citizen* and *The Charlotte Observer*.

But readers of the Observer weren't told that its author, Tom Wicker, was an employee of the state Department of Public Welfare - the byline read simply, "Special to the Observer." Wicker would soon become a reporter for the *Winston-Salem Journal*, and then go on to a long career as a national correspondent for *The New York Times*.

"I wrote, in effect, press releases - and hoped for the best. I didn't make any distinction in my own mind between the eugenics program and feeding the hungry," Wicker said. "I feel very badly about it in retrospect."

Many newspapers of that era believed their role was to help government officials, he said. "I think it was particularly true of journalists then. We were all kind of convinced that what our government was doing was right - that it wouldn't lie to you," said Wicker, who agreed there was a nave faith in science, too.

"It was a common belief that the whole world would soon be powered by cheap nuclear power," he said.

At a 1950 meeting of the Eugenics Board it was noted that "Mr. O'Keefe, City Editor for the *News & Observer*, said his paper did not use the Wicker story because of using a similar one the previous week ... He stated that they would be glad to use another feature story."

Newspapers were eager to use the copy, but the leaders of the Human Betterment League were shrewd enough to know that the fires had to be stoked. Minutes from a 1948 meeting contain a suggestion to send "a letter to each newspaper published in the state, giving estimated feebleminded born in county and suggesting sterilization."

The idea was approved, and in 1950 the minutes note that the group "has submitted 15 articles to 150 newspapers in the state."

In the 1960s, H.C. Bradshaw, the editorial page editor of the Durham Morning Herald joined the Human Betterment League.

Even as problems with sterilization programs began to come to light after a series of lawsuits in the early 1970s, readers in Winston-Salem were reassured that all was well.

"A case like the one in Alabama in which two young girls were allegedly sterilized without their parents' consent could not happen in North Carolina, the executive secretary of the state Eugenics Commission said Friday," read the opening of a 1973 *Sentinel* story.

But the *Journal's* review of the records of the eugenics board shows that children were sterilized despite the objections of their parents.

The "Lucky" Morons

Dr. Clarence Gamble wrote this untitled poem about 1947. He submitted it to the North Carolina Mental Hygiene Society in hopes that it would be used in pamphlets and other material circulated to drum up support for the state's eugenic sterilization program. He was surprised when the society declined to do so.

>Once there was a MORON, that means
>a person that wasn't very bright.
>he couldn't add figures
>or make change
>or do many things
>an ordinary man does.
>So he couldn't find a job
>and the RELIEF OFFICE
>had to help him out
>for YEARS AND YEARS.
>And one day he met
>another MORON
>who wasn't any cleverer than he was.
>But SHE was nicer to him
>than anyone had ever been.
>And so he MARRIED HER.
>And soon there was a BABY,
>and then ANOTHER
>and ANOTHER
>and ANOTHER.

And the welfare department
had to pay the family
MORE of the TAXPAYER'S
MONEY
and MORE
and MORE
and MORE
And when the children grew
up and went to school
They couldn't learn
very fast
because they had inherited poor minds from their parents.
They had to repeat MANY
GRADES in the school,
and never learned very much
and never were able to
GET A JOB.
and they cost the schoolboard
and the relief office
and the taxpayer
THOUSANDS OF DOLLARS.
AND THESE CHILDREN MARRIED
TOO - - -
So the story goes on
to grandchildren
and great-grandchildren
and so on forevermore.
Now there was another MORON
who also was a little stupid
and couldn't learn very
much but he lived in
NORTH CAROLINA

and that was very fortunate
for him.
For the Department of Welfare
in his county
Made him one of the
lucky morons
who went to CASWELL TRAINING
SCHOOL.
There he had a mental test
and he was taught a trade
simple enough to fit his brains,
and because the tests showed
he wouldn't ever be very
bright
Or be able to earn enough
to feed a family,
and because his children
might be feebleminded, too,
a surgeon performed
A SIMPLE OPERATION
which didn't change him AT ALL,
or take ANYTHING out of his
body, but kept him from
having any children.
And after a year or two
a JOB was found for him
which, because of his special training
he DID WELL,
and he earned enough
to be SELF-SUPPORTING.
And after a while he met a
GIRL

She, too, wasn't very bright,
but they liked each other.
And she, too, had been to
CASWELL for training
and had a JOB and a
surgeon had PROTECTED her from UNWANTED
CHILDREN, without
making her different in any other way from other women.
And because they loved
each other, they married
and WERE HAPPY just as other couples are.
Both kept on with their
Jobs so they were still
SELF SUPPORTING.
And there weren't any children's
mouths to feed ---- although
they wouldn't have
known why if
the operation hadn't
been explained to them.
And with just the two in the
Family, they kept on
being SELF SUPPORTING,
and they were very thankful they lived in NORTH CAROLINA.
And the WELFARE DEPARTMENT
DIDN'T have to feed them
and the SCHOOLS didn't
have to waste their efforts on
any of their children who weren't very bright.
And because they had been
STERILIZED, the taxpayers of

North Carolina had
saved
THOUSANDS OF DOLLARS
and the North Carolina MORONS LIVED
HAPPILY EVER AFTER.

Source: N.C. State Archives

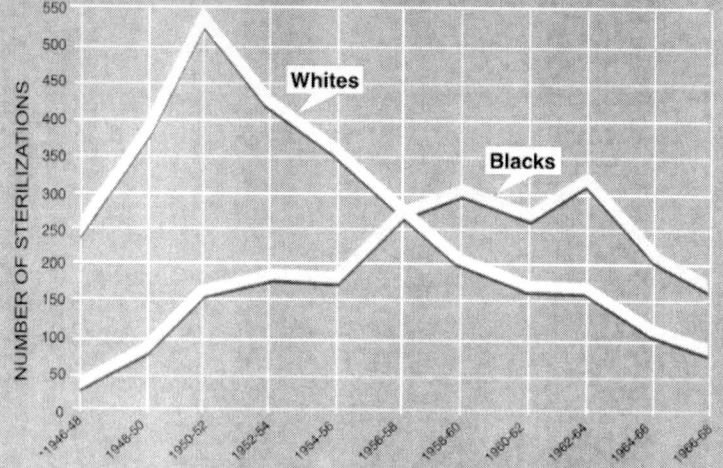

'Wicked silence'

State board began targeting blacks,
but few noticed or seemed to care about program

By John Railey
JOURNAL REPORTER

State Sen. Wilbur Jolly's voice thundered as he spoke at a public hearing in the legislature on April 1, 1959. Out-of-wedlock births were soaring and he had a solution. After an unmarried woman gave birth for the third time, he said, she should be sterilized.

A group of blacks was in the audience, and Jolly shook his finger at them as he made his points. "You should be concerned about this bill," said Jolly, a white Democrat from Franklin County. "One out of four of your race is illegitimate."

The blacks stood up and demanded to be heard, but in a state and legislature controlled by whites they were ruled out of order.

Jolly's bill never made it to the Senate floor, but in a sense it didn't matter. For years, the Eugenics Board of North Carolina had been quietly ordering sterilizations for unwed mothers. And beginning in the mid-1950s, the board had begun a pattern of sterilizing a disproportionate share of blacks.

"I relate that whole way of thinking to a way that Hitler would think during Nazi Germany," said Tony Riddick of Winfall, whose mother was sterilized when she was 14.

The sterilization program happened in a state whose leaders - then and now - congratulated themselves on their progressive attitude on race relations. In 1960, the year after Jolly introduced his bill, students from N.C. A&T State University would begin the sit-in movement and a moderate named Terry Sanford would beat back a segregationist challenge and become governor. And yet in 1960-62, the eugenics board ordered 467 sterilizations – 284 were for blacks. This in a state where minorities were only a quarter of the population.

Now, as the details of the board's work - about 7,600 sterilizations over more than 40 years – are becoming public, many activists are outraged. "It's clearly genocide," said the Rev. John Mendez of Winston-Salem, the president of the Ministers' Conference of Winston-Salem and Vicinity. "Genocide is the last stage of racism; it's where you start exterminating people when all other things fail ... just to make room for your race and your race only."

The Rev. Jerry Drayton, another local civil-rights activist, said he was frustrated by revelations about the sterilization program. "It's so much unfairness in the behavior of human beings that certain things just don't surprise me," he said.

Tom Lambeth of Winston-Salem, the administrative assistant to Sanford in the early 1960s, said he didn't even hear about the eugenics board until he was told about it by the *Winston-Salem Journal*. "Using a medical procedure for any racially motivated purpose, I don't think you can defend that in any way. A racial bias is what we should be ashamed of. And maybe we didn't do it as much as others, but I have no idea," Lambeth said.

Jolly, 86, is still practicing law in Louisburg. He said he has no regrets about his bill. "I doubt very seriously it was constitutional, but something needs to be done to prevent the burden of there's certain people who just can't wait to have a child in order to get a check so they won't have to work, they live off welfare," he said recently.

The Rev. W.W. Finlator of Raleigh attended the hearing on the Jolly bill. He was one of the first whites to join blacks in the civil-rights movement, and he now regrets not having taken a stand against sterilization that day or later.

"There's a phrase that Martin Luther called 'wicked silence,'" said Finlator, 89. "There was a time when I was guilty of wicked silence."

Finlator may not have been alone.

Loaded issues

Race and procreation have always been loaded issues in the South. Throughout slavery, whites encouraged their slaves to reproduce to create more bodies to work and sell. But by the 1950s, some whites were fretting about supporting blacks through welfare.

During that time, the numbers of black girls and women ordered sterilized by the eugenics board increased. An emphasis on targeting poor blacks for sterilization seeped into the state's network of welfare departments. According to records from the eugenics board, few of the heads of these agencies objected to the eugenics program, and many bought into it.

"These superintendents were neither petty autocrats nor fanatics, but they generally agreed on the value of eugenic sterilization in reducing both general relief and ADC (Aid for Dependent Children) payments," Joseph L. Morrison wrote in a 1965 article in The Social Service Review, a trade magazine for social workers.

"Since Negroes accounted for a disproportionate share of illegitimate births 'subsidized' by ADC, the racial aspect of the superintendents' intent was clear enough," wrote Morrison, a faculty member at the journalism school at the University of North Carolina at Chapel Hill.

Some blacks remember women being threatened with the loss of their welfare benefits if they didn't submit to sterilization.

"When basically your only income is welfare, your livelihood is welfare, they (women) did what they had to do. And that's sad," said the Rev. Lee Faye Mack, 70, of Winston-Salem.

The sterilization program was whispered about in black communities, but evidence of its racial bias was not publicly scrutinized at the time. There is little if any record of Sanford - revered for his work in race relations - even mentioning the program, and Lambeth doubts that Sanford realized the turn the board had taken toward sterilizing disproportionate numbers of blacks.

"Who knew?" Lambeth said. "My guess - and unfortunately he's not here to ask - is that it never came across his desk."

Newspaper editorials and stories praising the program left out details about children and their parents fighting sterilization orders in vain. They left out references to people of normal intelligence being termed "feebleminded" as a justification for sterilization. And they left out statistics showing that more blacks were being sterilized than whites.

Those factors and others, including the complicity of some blacks, allowed the work of the eugenics board to proceed with few obstacles. "They (some mothers) weren't taking care of their children like they should," said Lula Morrison of Winston-Salem, a black who worked as a nurse for the Forsyth County Health Department. "It had to be some way for them to stop having them."

"Anybody that keeps having children and can't stop having them, it's something wrong with them," said Morrison, 95.

Another black who worked as a nurse for the county, Ida Ruth Staplefoote, agreed, adding that that those sterilized were mentally handicapped. "We just didn't want a lot of unwanted, uncared-for children," said Staplefoote, 91, of Winston-Salem.

"Of course, it's a different ballgame now. Some of them have as many as they want to."

Black doctors performed some of the sterilizations, the nurses noted.

When blacks targeted for sterilization resisted, they were often subjected to intense pressure. In Sterilization in North Carolina, published in 1950, Moya Woodside wrote that nurses and social workers found that "the Negroes tend to resist more than the whites which may be due to ignorance and superstition which is more prevalent among the Negroes than the whites. The very ignorant of both races are very suspicious of this type of operation."

While at Shaw University in Raleigh from 1968 to 1972, Mendez said, he and other student activists tried to educate blacks across the state about issues including the threat of sterilization. But they lacked details about the program, he said.

Even if more details had been known, it is doubtful that this issue of women's rights would have been a high priority in a civil-rights movement led by men who already had a long list of demands.

"I guess it was so many other things going on," said the Rev. Juanita Tatum of Winston-Salem. "During the '60s, we were trying to get our foot in the door to sit down at a restaurant or to get a job."

Weighing the costs

The 1959 hearing on Jolly's bill was one of the few points where the issue of sterilizations erupted into public debate. The bill - and a similar one in the House sponsored by Rep. Rachel Davis, a Democrat from Lenoir County - would have mandated that a woman who had more than two children out of wedlock prove that she was not "grossly sexually delinquent" or face sterilization by order of the eugenics board. The health committees of both the House and the Senate held the public hearing.

Jolly said recently that he doesn't regret his outburst at the blacks. "It wasn't the proper thing to say probably politically, but it was the truth. Even now."

But his bill wasn't racist, he said. He said he didn't know that the eugenics board was already ordering the sterilizations of women who had children out of wedlock, and said that more action was needed on the issue of illegitimate births. "One (generation) right on top of another, and they use this as a means of livelihood."

Finlator said that money shouldn't be a consideration. "This guy Jolly and others were thinking about the costs of all these little black babies coming into the world. They were thinking not so much about the families of these people, but how much it was going to cost."

The legislature finally disbanded the eugenics board in 1974, 15 years after Jolly pushed his bill. "Sterilization is a vicious atrocity - gruesome, cruel, dehumanizing, degrading, barbaric, unethical, un-Christian and unconstitutional," Rep. Joy Johnson, a black preacher from Robeson

County, told a House committee shortly before the end of the eugenics board.

By then, the culture of the South and the nation was changing as blacks and women were being heard more and more. They were beginning to share in power, although respect didn't always come with that. Prejudice lingered, and many activists say that it lingers to this day.

"The mistreatment of people is an attitudinal thing, and that attitude has not changed," Drayton said.

DETOUR: In '48 state singled out delinquent boys

By John Railey and Kevin Begos
JOURNAL REPORTERS

The seven boys are old men now, if they're all still alive. They are senior citizens who may well carry haunting memories about what happened to them while they were at the Stonewall Jackson Training School in 1948. Most of them were all set to be released, but before they could leave the school outside Concord, the state of North Carolina wanted to sterilize them.

Harry Truman was president and the thousands of veterans who had fought in World War II had returned home and were pushing the United States toward its boom years. There was order in the land, and those who trespassed against that order were dealt with firmly in a time when neither capital punishment nor corporal punishment were subject to debate.

Across North Carolina, boys and girls who broke the law - as well as ones who were simply promiscuous or truant - were sent to reform schools such as Jackson. The fat folders on these schools at the state archives in Raleigh are full of yellowed letters in which state officials praise each other for their efforts to help these children. Tucked away amid all those images of backslapping are a few pages that tell grimmer stories, ones of children crying unheard, lost in the shadows of a system where being beaten and locked in closets was not all that unusual.

During a 1938 hearing before the Eugenics Board of North Carolina, a 13-year-old girl was asked why she wanted to go home from a county institution.

Patient: 'Cause I don't want to stay up in the lockup all time.
Mrs. Bost: What do you mean, lockup?
Patient: They call it bread and water room.....
Dr. Stimpson: They keep you locked up in this room.
Patient: Yes.

For hundreds of youth among the thousands in North Carolina institutions, being sterilized by the state was also a fact of life. For these young people, who were categorized as "feebleminded," the operation was a prerequisite for release. The eugenics board urged it, and the social workers the board kept in contact with often insisted that institutionalized students be sterilized before returning to their counties.

"We would say this person is ready to return to the community ... the community would say, 'No, we can't accept her back because she'll get pregnant and there'll be another child on welfare," said Vernon Mangum, who headed the O'Berry Center, a training school in Goldsboro, from 1959 to 1978.

Only a few girls were sterilized at his school, he said, and they were all mentally retarded.

Asked if any students ever complained about the operations, Mangum said, "Not to me.

"I was at the top of a ladder. I would always see these people and review the cases and would talk with them, but I never approached the subject itself with them. This was a legal type of thing, and I was not about to get involved with the legal part of it, other than following the laws of North Carolina."

Mangum said that no boys were sterilized at O'Berry. "They didn't bear the children."

For the most part, boys in institutions were not sterilized, just as men in the general population were not forced to undergo the operation. But that was about to change at Stonewall Jackson, a boys institution in a farmland setting.

Changed lives

Boys were sent to Jackson for minor scrapes with authorities, not because of mental illness. Up until 1948, there had been no sterilizations of boys at Jackson, but that summer - for reasons unknown - the eugenics board targeted seven boys out of the approximately 300 at the school.

"Most of these boys are ready for discharge but cannot leave until action is taken on the authorization of the eugenics board," according to a memo in the school's file.

The sterilizations of the Jackson boys - whose names were blacked out in eugenics board records reviewed by the *Winston-Salem Journal* - were delayed.

The surgeon, according to Dr. King, hesitates to perform the operations fearing some of the boys might be psychotic and later return to Concord to do physical harm to him. – Memo from the eugenics board

There was a quick solution to that problem, though. The eugenics board authorized a different surgeon to perform the operations.

The first surgeon's reluctance was not the only factor holding up the operations. A memo suggested that a change in superintendents at Jackson may have slowed the process, as well as uncertainty about the mental evaluations. The eugenics board urged that a psychiatrist "visit the school at the earliest possible date."

The boys were soon deemed "feebleminded" by intelligence tests of the era, but some of the one-paragraph descriptions that the eugenics board voted on hardly seem to justify a sterilization operation.

Single boy, 15 years of age, who was admitted to the training school ... because of delinquency. He repeatedly stayed away from home a number of days at a time and would not obey his parents. His mother, who has epilepsy, feared leaving him alone with his sister. In the institution he has been found to be tempermental (sic), untruthful, and requires constant supervision.
– Eugenics board minutes, February 1948

Another 14-year-old, also sent to the school for delinquency:

Has a marked speech defect and for a period of time was a student at the State School for Deaf.... His family are known to a number of social agencies and have received intermittant (sic) help from DPW since 1936.
– Eugenics board minutes, February 1948

As the process moved forward, the consent of one father was "signed by mark," indicating he was illiterate. And along with the hot days of summer came more cryptic hints of trouble.

A visit was made to Stonewall Jackson Training School ... in an attempt to clarify misunderstandings regarding sterilization operations authorized in that institution. – Eugenics board minutes, June 1948

The records show that six vasectomies were carried out on Jackson students from July 1948 to June 1949 - the first and last such operations in the history of the institution. More details about what happened at the Jackson School may be under seal at the state archives, or they may have been lost.

Six boys had their lives changed forever, while one, for unknown reasons, managed to avoid the order sent down from Raleigh.

In the years ahead, the focus of the eugenics board would shift from girls and women in institutions to girls and women outside the walls of reform schools and hospitals, and the number of boys and men sterilized would continue to decline.

Just Carrying Out Orders

Some doctors performed sterilizations with no qualms; others, looking back, recall having some reservations

By John Railey
JOURNAL REPORTER

Doctors across the state quietly carried out the orders of the Eugenics Board of North Carolina through the Great Depression, World War II and the boom years that followed.

But in the face of professional criticism, some surgeons stopped performing the sterilizations on girls and women in the 1960s. "Most physicians who were involved in it began to say it wasn't worth it," said Dr. A.V. Blount, 80, of Greensboro. "They were being largely criticized by their colleagues."

Blount said that he assisted in sterilization operations and may have done one himself, although he can't remember. Either way, he said, he always had reservations about the operations. "To do this to the mentally ill, sterilizing them is a little bit harsh," said Blount, who is still practicing medicine part time. "I had a feeling that ... perhaps this wasn't a good thing to do."

Dr. Robert Albanese, a family-practice doctor who helped arrange a sterilization in Dare County in 1967, said he now has reservations about doing so. The eugenics program "seemed like a good idea," he said, "but then again to do a surgical procedure on somebody who doesn't really want it, that's not right, either."

Most of the doctors who did the sterilizations are dead, and few of the surviving doctors will talk publicly about the issue, saying that it is still too sensitive.

But one doctor who performed the operations said he felt that he was helping society. "I never did any that I didn't evaluate and think that they were incompetent of having a child," said Dr. Ernest Brown, 71, of Lumberton.

Other doctors who declined to comment for this story minimized their roles, saying that they only carried out state orders. As eugenics board members and their supporters pushed the program, it was usually social workers - not doctors - who took the initiative in sterilization cases.

"With a few exceptions, the evidence at our disposal suggests that doctors do not interest themselves in the possibility, and rarely trouble to bring cases of feeble-mindedness among their patients to the attention of the Eugenics Board," wrote researcher Moya Woodside in Sterilization in North Carolina, a book published in 1950. She wrote that "medical indifference towards sterilization was mentioned not only by social workers but by doctors in the public-health service who criticized their colleagues as lacking in social-mindedness."

Brown said that mothers of his patients would often ask the state to order the operations. He didn't do more than 12 of the operations, he said, and all of them were during the 1960s.

"Most of them (the patients) were in their late teens up into the 30s. These children didn't know what they were going through, they would provoke unscrupulous men, if you know what I mean," Brown said.

The patients, of low intelligence, never voiced objections, he said. "I don't think the patients ever understood what was happening."

The eugenics program was the best thing the state could come up with at the time, he said. But he added that "we haven't lost anything by not having the state do it."

Blount said that the organizers of the eugenics program had good intentions. "I think honestly, the people who were involved in it thought they were helping."

Church Silent
Scarce Catholics no threat to N.C. drive to sterilize

By John Railey
JOURNAL REPORTER

The Rev. Martin Collins of Winston-Salem sat down to write a letter in March 1948. The words he wrote would stand as one of the few religious protests against the sterilizations pushed by the Eugenics Board of North Carolina.

"Sterilization is immoral in itself and contrary to God's law, be it either voluntary or imposed," Collins wrote in the letter, which ran in the *Winston-Salem Journal*. "The state has no more right to mutilate individuals when they have done no wrong than individuals have to mutilate themselves."

The letter from Collins, the priest at tiny St. Benedict the Moor Catholic Church, had little effect. The eugenics board would approve thousands of sterilization petitions during the next 25 years. And only now, as the details of its work emerge, are many others sharing Collins' outrage.

"I'm not surprised, but I'm appalled," said Irma Gadson, a member of St. Benedict's since the time Collins served there. "But I wouldn't be surprised at anything they did, because I'm an African-American or whatever you want to call me."

It's hard to believe in these times of loud and public debates about abortion, but Collins was virtually alone in his public outrage. The priest was one of a handful of Christians to speak out against a statewide program that moved quietly and efficiently. The early proponents of the program anticipated this lack of opposition from the pulpits and saw it as a plus.

"North Carolina also presents advantages for genetic study and the application of eugenic programs in that there is practically no public opposition in this area to such programs," wrote Dr. C. Nash Herndon, a professor in the Department of Medical Genetics at the Bowman Gray School of Medicine, in the 1940s. "No church opposition has yet been encountered to proposals of family limitation or sterilization, and in fact collective support has been obtained from ministers of various denominations."

In the years ahead, proponents of the program carefully took note of opposition from people of faith. As they considered Lumbee Indians, they noted that some of the "Holiness" followers among that group of American Indians were against sterilizations. As they studied groups of blacks and whites around the state, they noted opposition from "fundamentalists."

"Indeed, what might be called 'the-Lord's-will-point-of-view,' with an associated fatalism and clear-cut notions of 'right' and 'wrong' is common to a large part of the rural population and is especially marked among Negroes, for whom religion assumes an intensely personal and emotional character," Moya Woodside wrote in Sterilization in North Carolina, published in 1950.

"The Bible was prominently displayed in many white and Negro homes which were visited, and women who had been sterilized often reported prolonged heart-searchings before they agreed to the operation," she wrote.

Jews knew all too well of the sterilizations in Hitler's Germany, but they were only a tiny minority in North Carolina. Even so, Woodside described reports that some Jewish physicians "would never sign a sterilization petition, however commendable the case."

Conservative Protestants, including those whose wives and daughters were targeted for sterilization, occasionally spoke out, but there were no Jerry Falwells or Ralph Reeds to organize them. Black Protestants would soon band together in the civil-rights movement, but reproductive rights were not on their list of issues.

And some Protestant leaders, starting with William Louis Poteat of Wake Forest College, pushed eugenics. "We have seen the peril of feeble-mindedness and insanity multiplying under the cloak of silence... Probably 8 percent of us are a burden on the back of the rest of us," he told a group of Baptist educators in 1921, according to Randal Hall's biography of Poteat, a legendary Wake president.

Most of the opposition to the eugenics movement came from Catholics. As the eugenics movement was cranking up in Germany and across this country in the 1930s, the Catholic Church was already taking a firm stand against sterilization - just as firm a stand as it now takes against abortion.

In the late 1940s, there were only about 10,000 Catholics in North Carolina, a state that had millions of Protestants. And only a few in the Catholic minority were speaking out in a time when they faced prejudice for their faith alone.

"However, it was clear that from a practical point of view, organized Roman Catholic opposition, the greatest hindrance to sterilization in many other states, was negligible," Woodside wrote. "Whatever Catholics individually may believe, in North Carolina their numbers are too small for them to have serious influence on the Legislature, nor can they impede the sterilization program except by individual refusals to cooperate - as for instance in the case of surgeons or certifying physicians - or by denying hospital facilities to sterilization cases."

They could also write letters.

Which is what a few Catholics, including Collins, did. He was a Franciscan priest who served here from 1943 to 1952. He was a white priest at a poor black parish, a common arrangement in the Catholic Church but yet unheard of for most Protestants.

It is unclear whether he acted from his faith alone, or whether one of his parishioners had been targeted for sterilization or sterilized.

What is known are his words:

"Behind this idea of sterilization is the implied supposition that only those of social standing, economic complacency and with a certain amount of formal education should be allowed to propagate.... One thing is certain and that is that the birth controllers will be punished, either in this life or the next. Perhaps one form of that punishment will be for them to see the poor and the 'feeble-minded' inherit the earth."

Sterilization was often the way out

State hospitals, training schools had captive population, but staffers and social workers were sometimes at odds

By Kevin Begos
JOURNAL REPORTER

She was 17, just a girl, and she was in the State Home and Industrial School for Girls in Moore County. The state was considering releasing "Peggy," but there was a condition. She would have to be sterilized first.

At a hearing before the Eugenics Board of North Carolina in 1938, Peggy's mother pleaded for her daughter's future.

"If she is not operated on, won't she ever come home?" her mother asked.

"I don't know," replied Paul Mickey, a member of the eugenics board.

"I object on grounds of 'feeblemindedness,'" Peggy's grandfather said. "She was not feebleminded when she left home. I can get plenty of witnesses to say that she was not."

"The main thing is to keep feeblemindedness from becoming a great problem," said Dr. J.C. Knox.

Mickey added that "the state has placed (the) decision in the hands of government agents, and we can't go beyond that."

"I object and if you go ahead and do it and she doesn't get along all right, I am going to sue the state," Peggy's mother said.

The eugenics board approved the sterilization.

Institutions in North Carolina made extensive use of the state's eugenic sterilization law in its early years, and they frequently made sterilization a condition of release. The 1935-36 biennial report of the eugenics board notes that "none of the inmates of Caswell Training

School should be released before being sterilized, except in the few instances where normal children have been committed through error."

All the cases that came before the eugenics board were supposed to include some sort of "consent" by parents, patients or guardians, but there was a catch. In the 1930s and 1940s, the eugenics board often refused to release patients from institutions before sterilization.

Institutional sterilizations peaked at 152 during the 1938 reporting year and then started to decrease. In 1947, there were just 65. About a third of the 7,600 sterilizations authorized by the state were for patients in mental institutions and training schools.

In the late 1940s, the Human Betterment League of North Carolina had started a statewide publicity campaign to promote an expanded sterilization program.

But the psychiatrists and mental-health professionals who knew most about the issue proved to be the toughest audience, and this skirmishing continued for years.

In 1954, the Human Betterment League met with Dr. Ellen Winston, the head of the N.C. Department of Public Welfare and a eugenics board member.

The minutes of the meeting note that "the large decrease in the number of institutional sterilizations for the calendar year 1953. Winston cited the fact that some superintendents and staff members of state hospitals are unsympathetic toward sterilization."

That was because most people in the mental-health field were aware that IQ tests were flawed and that the entire concept that genes always passed on intelligence was wrong, too, said Gilbert Gottlieb, a research professor of psychology at the University of North Carolina at Chapel Hill who worked at Dorothea Dix Hospital in the late 1950s.

Gottlieb recalls discussing the sterilization program with other professionals in the late 1950s. Dr. Eugene Hargrove was the state mental-health director and a member of the eugenics board in the early 1960s, and he was "unhappy" about the sterilization program, Gottlieb said.

Sterilizations in institutions

State hospitals

	Total	Female	Male
Broughton	829	225	604
Dix	632	98	534
Cherry	487	191	296
Umstead (1)	34	7	27

Schools for mentally defective

	Total	Female	Male
Caswell	597	252	345
Murdoch (2)	14	0	14
O'Berry (3)	12	0	12

1 - Opened in 1947
2 - Opened in 1958
3 - Opened in 1957
4 - Name changed to Dobbs School for Girls when it moved to Kinston in 1947

Source: Eugenics Board of North Carolina

Broughton Hospital (Morganton); Murdoch Center (Butner); Umstead Hospital (Butner); Dorothea Dix Hospital (Raleigh); Eastern Carolina Training School (Rocky Mount); O'Berry Center (Goldsboro); Cherry Hospital (Goldsboro); Stonewall Jackson Training School (Concord); State Industrial School for Girls (Eagle Springs); Caswell Center (Kinston); Dobbs School for Girls (Kinston)

Training schools

	Total
State Home and Industrial School for Girls (white, female)	300
Stonewall Jackson Training School (white, male)	6
Eastern Carolina Training School (white, male)	4
State Training School for Negro Girls (4) (black, female)	3

Eugenics board records reviewed by the *Winston-Salem Journal* show that Hargrove cast an increasing number of "no" votes on sterilization cases as the 1960s progressed.

But if mental-health professionals had become wary of sterilization, the *Journal's* review of eugenics board records shows that social workers continued to push the program.

> *Because she has gotten along so well at Caswell, the staff now feels that she is able to return to the community. However, before she is returned to the community due to her sexual promiscuity the Department of Social Services feels that she needs sterilization.*
> — Eugenics board minutes, 1969

It's unknown whether that petition was approved.

In all, there were 597 sterilization operations performed at Caswell from 1929 to 1968. At Broughton Hospital there were 829; Cherry Hospital, 487; Dorothea Dix Hospital, 632; and John Umstead Hospital, 34. Over the life of the program, at least 2,900 institutional sterilizations took place.

The eugenic sterilization program is gone, but mental-health professionals say that some of the ethical issues about how to treat people with disabilities remain.

"We serve about 400,000 people in (public) mental-health centers every year, and I do think we have come light years in terms of civil and legal rights," said John Tote, the executive director of the Mental Health Association of North Carolina.

"But who does have access to the best, the newest, the most effective treatments? As you have (mental health) decisions made more and more on cost, you bring up the whole (ethical) issue," he said.

Tote said that a core flaw of the eugenics movement was the idea that society, or people, could be made flawless.

"Quite frankly, none of us are (perfect)," he said. "We will never have a perfect society in terms of physical issues. Just because somebody is in a wheelchair or on crutches or has schizophrenia, doesn't make them any less of a person."

Treating those who are different from some mythic ideal isn't just a question of compassion, he said.

"Part of it is, we will be judged on how we treat those folks," Tote said. "Society can also be enriched by how it responds to the needs to others."

'BAD' GIRLS: Indians posed a tricky race problem for the state

By John Railey
JOURNAL REPORTER

Robeson County officials made a seemingly simple request in 1938. They wanted to send two girls they had labeled as "delinquent" to a state training school.

But there was one big problem - the girls were Lumbee Indians, then segregated from whites and blacks and kept out of most state training schools. The question about whether they should be admitted festered into arguments about racial purity and Indian attitudes toward sterilization, operations that often happened at the state schools. Almost 65 years later, the admissions controversy sheds light on the tension surrounding the eugenics movement, the sterilizations that resulted from it and the state program's preoccupation with race.

It is emotion-laden terrain, particularly so for American Indians. More than 50 Indians were sterilized in North Carolina, a small portion of thousands sterilized nationwide as the eugenics movement kicked in during the first half of the 20th century. Today, many American Indians freely speak out about that time, seeing it as part of a pattern of physical and cultural genocide that they long faced. But in 1938, whites often did the talking for them.

"We have failed to secure any evidence that the Indians have hope of preserving their own race as a distinct one or of preserving any item of their culture," psychologist Harry Bice wrote in a report to state officials about the admissions controversy over the Lumbee girls.

The girls, ages 14 and 15, were being considered for admission to Samarcand Manor at Eagle Springs, also known as the State Home and

Industrial School for Girls. White workers with the Robeson County Department of Public Welfare said they were sexually active and of borderline intelligence, according to state records. At the time, blacks were still barred from training schools for whites. A 1933 state law required that Indians be admitted to Samarcand and a training school for boys, but law and practice were separated by inevitable foot dragging.

'The unfit should be sterilized'

In an evaluation of the 14-year-old, Bice wrote that "since the girl is mentally deficient and persistent in delinquency, she should be sterilized." From the 1930s through the '60s, hundreds of girls and several boys were sterilized at state training schools on the basis of similar reports. The orders for sterilization came from the Eugenics Board of North Carolina, which the legislature formed in 1933.

In his report to the state, Bice quoted a few Indian leaders from Robeson County. "If Indian girls at Samarcand were sterilized, it would be a good thing - the unfit should be sterilized," a public-school principal identified only as "Mr. Lowry" said. "The lower class and the Holiness people would fight it, but there is no teaching of the Indians as a race to oppose sterilization."

Much of the admissions controversy dealt with questions about the purity of the Lumbees, who were once called "The Cherokee Indians of Robeson County." Glenn Ellen Starr Stilling of Appalachian State University, the author of The Lumbees: An Annotated Bibliography, notes that eugenicist Arthur Estabrook of the Carnegie Institution of Washington studied the Lumbees. In the 1920s, Estabrook co-wrote a thinly veiled account of the Lumbees as an example of a race that had become impure by mixing with blacks.

In a chapter of Mongrel Virginians, Estabrook and Ivan McDougle called Robeson County "Robin County, N.C.," and called the Lumbees the "Rivers" tribe.

As Bice prepared his report in 1938, he touched on similar issues. He wrote that the 14-year-old he had recommended for sterilization had a broad nose and dark color. "Whatever the race of this girl may actually be, she would not be accepted by the girls at Samarcand, nor by white people in general, as anything but a Negro.... She is reported to have kept company with a boy that is recognized as a Negro."

Throughout the history of the eugenics program in North Carolina and across the country, evidence of interracial relationships was often a justification for sterilization.

Today, the obsession that whites had with the Lumbees and race seems bizarre to some Lumbee leaders. "It does appear to be folly, less than petty, insane almost," said the Rev. Mike Cummings of Robeson County. "But that marked our lives 50 or 60 years ago. And it does linger."

The Lumbees have fought for years to have the federal government recognize them as a tribe. Some Lumbees are still prejudiced toward blacks and don't acknowledge their own black blood, Cummings said. But what defines them as a people more than bloodlines is a long history of shared experience, he said.

Boiling the definition down to blood carries the threat of erasing the Lumbee people's history as distinct group, Cummings said.

The Lumbees, lacking tribal status, never had a reservation like the Cherokees have in Western North Carolina. In the early 1950s, the eugenics board explored the possibility of ordering sterilizations on some of the residents of the reservation. A lawyer with the attorney general's office advised the board that it lacked authority on the reservation.

Before that advice was given, some sterilizations were performed on the reservation, according to eugenics board records, but it is unknown whether those operations were by order of the board. Board records do not indicate the tribes of the Indians sterilized in North Carolina.

Against protest

Ultimately, Bice wrote that the 14-year-old should not go to Samarcand. Instead, he wrote, she should be sent to a particular family's home to be trained as a domestic worker. "In color, they are much like the girl," he wrote.

The records don't indicate whether the girl was ever sterilized. Nor do they indicate if the 15-year-old underwent the operation. But the older girl was sent to Samarcand.

"We will admit this girl, against protest, believing that it will be hard for her to make proper adjustment in a school with white girls," the school superintendent, Grace Robson, wrote in a letter to Robeson social workers.

Tempting Choices
The new frontier must not forget
to ask why, who, geneticists warn

By Danielle Deaver
JOURNAL REPORTER

Although government involvement in sterilization has ended for the most part, the chances of another movement that wants to use science to eliminate human flaws are increasing, experts say.

The unraveling of the human genome opened a new world of possibilities for people who think that they can engineer a better human race.

Eugenics never really goes away, because some things about human nature never change, said Arthur Caplan, the director of the Center for Bioethics at the University of Pennsylvania.

"People always say to me, "Why do people want to do this?' Well, here's a moral principle — make your children's lives better than yours. If you believe it, then you're right toward the idea that you should do this," Caplan said.

But not everyone agrees that genetically engineering people is a good idea.

"There's this tendency to want to bring into the world only genetically perfect babies, and to eliminate (imperfect) babies. And there is a possibility of a eugenics movement that would be not racial, but promoted by insurance companies, genetic counselors, genetic screening," said Barry Mehler, a professor of history and the director of the Institute for the Study of Academic Racism at Ferris State University in Big Rapids, Mich. "And the rationale for that would be to bring healthy babies into the world. Who can argue against that?"

We're doing it now

The question of how much people want healthy babies is answered every day in the offices of genetic counselors.

Pregnant women can undergo screening and diagnostic tests for a multitude of potential problems — Down syndrome, spina bifida and other diseases that could leave the baby mentally retarded, sick or fatally injured.

There are only a few prenatal treatments for the illnesses that can be detected in the womb, but the tests offer other information that will help parents, said Daragh Conrad, a certified genetic counselor at the Wake Forest University School of Medicine. Knowing the extent of the illness can help parents and doctors prepare for the delivery of a sick baby and can let parents get ready for the challenges of raising a child with a disability.

But before parents can get to that point, they first have to make another decision — whether to have the baby or terminate the pregnancy.

At Conrad's office, fewer than half of the parents who find out that they are having a baby with Down syndrome or spina bifida choose to abort. More than half of the parents who are having a baby with more severe problems opt for abortion, she said.

Prenatal testing and its potential to lead to abortion troubles some people.

"There are several concerns. One concern about prenatal testing is a religious one, for people who believe that life begins at conception and fetuses have equal rights," said Dianne Bartels, the associate director of the Center for Bioethics at the University of Minnesota.

Advocates for the disabled are also troubled by the idea of aborting less-than-perfect babies. The idea came under new scrutiny a year ago when the American College of Obstetricians and Gynecologists recommended that all prospective mothers be offered prenatal testing to look for the gene that causes cystic fibrosis. There is no treatment right now for cystic fibrosis that can be done before birth.

Cystic fibrosis is a genetic disease that affects 30,000 people in the United States. It causes thick, sticky mucous to be produced, which then affects the lungs and other organs. Most children with the disease live past age 20; about 25 percent live past age 35.

"Screening for cystic fibrosis means that we are screening for a disease that may affect the quality of a person's life, but it's not uniformly fatal. So we are doing like the eugenicists did, perhaps we are deciding what lives are worth living," said Dr. Anne Drapkin Lyerly, an assistant professor of obstetrics and gynecology and a faculty associate in the Center for the Study of Medical Ethics and Humanity at Duke University.

But ethicists in general also have problems with people who abort fetuses for reasons other than health concerns.

"For some people, it is not wrong to terminate a boy or a girl if for instance you have too many of that gender already. And in some cultures, having a boy is so important that it is routine to do genetic testing before," Bartels said. "Many of us are concerned about how and why you can draw that line between health and attributes."

Aside from the larger issues of medical ethics, aborting fetuses for reasons of gender can lead to societal problems, such as the ones in China and India, where the gender ratio has become skewed because of selective abortions.

In some areas, the problem has become so pronounced that men are having trouble finding wives.

Building a better baby

As science advances, it offers more choices for prospective parents.

Adam Nash was an unknowing pioneer when he was born in Colorado on Aug. 29, 2000 — the first baby born using both in vitro fertilization and a technology called preimplantation genetic diagnosing.

Adam's sister Molly, then 6, had a rare disease, Fanconi anemia, that could lead to leukemia and death. Her parents wanted to have another child who would both be free of the disease and a stem-cell match for his sister.

Doctors screened 10 embryos produced using the Nash parents' eggs and sperm, and kept the one that was free of disease and a stem-cell match.

Adam was born healthy, and his sister had a stem-cell transplant five weeks later.

The technology used to produce Adam is likely to be used more in the future. Eventually, when genetics is better understood, the pre-implantation testing, or other, more advanced technology could be used to find embryos that have some of the attributes that parents want — almost custom-made babies.

For some people, it's a logical next step.

"There's already shaping of kids. Some of it is innocuous. As long as it's not harmful, we can do that. And the likelihood is some genetic tests might be very attractive, like genetic engineering to give us perfect vision. It might put the laser surgery out of business," Caplan said. "It's hard to see what the downside is of having better vision, lower blood pressure or whatever. As long as you understand you don't have to."

But some wonder if this is the proper use of the technology.

"I think the concerns from health professionals are that we'd like to believe that health technologies will be used to further the health of the child, and we're very skeptical as a general rule of things that would actually be enhancements," Bartels said. "We're not here to give people whatever would make them happy in other ways."

Even the Nash situation was troubling to some people, Bartels said, because although no one objected to selecting a healthy embryo, there was concern about selecting an embryo for a trait that had nothing to do with the baby's health, but the health of his sister.

There are also fears about creating children to fit their parents' expectations.

Now and then...

1929

"Eugenic sterilization is no panacea, but it is one of the many tested and dependable measures that will help reduce the burdens and increase the happiness and prosperity of the population in this and future generations. ... If recognized as an integral part of a broad system of protection and supervision of those unable to meet the unaided the responsibilities of citizenship in a highly competitive industrial system, it can be productive only of good."
— *Human Betterment Foundation*
Pamphlet promoting the use of eugenic sterilization

2001

"Genetic testing and genetic information is but one of many new technologies that will advance health care, provide better preventive medicine and counseling, unlock the causes and factors in diseases, make for better treatments and improve the delivery of services. We can ensure that this new technology is a friend to patients and not something that they, in any way, need to fear."
— *U.S. Rep. Cliff Stearns, R-Florida*
Congressional hearing on the potential for discrimination based on genetic testing

"People are terrified of creating a genetic underclass."
— *Arthur Caplan*
Director of the Center for Bioethics at the University of Pennsylvania

"Going into a pregnancy thinking that you can choose the type of child you can have is basically not the right attitude that someone should enter pregnancy with. We hope parents will love whatever child they end up with. There will always be some things we can screen for, but if you want to get your "blue-eyed, blond hair' child that you want, if they get leukemia when they're 3, you still have to love them and be their parent, even if they're not going to fulfill whatever dreams you had," Lyerly said. "Parents have to be able to accept their children for what they are."

Economic pressure

Also troubling to ethicists is the possibility that genetic engineering could become so widespread that it would become necessary to engineer children in the future just to make them competitive in society.

"You could see shifts in the make-up of the population that are not good for the population. That's possible. The other thing that could happen ethically is you could not fix the equity problem between the rich and the poor, so that the poor will have even worse health than they do now," Caplan said. "People are terrified of creating a genetic underclass."

Economics could be a factor in genetic engineering in other ways. People worry that insurance companies could insist on prenatal testing and then refuse coverage for children whose birth defects are discovered before birth. Many states forbid health-insurance companies from using genetic information when deciding whom to insure or how to insure them, but the protection is not total. North Carolina law forbids companies from refusing to insure people with the sickle cell trait or hemoglobin C trait, or from charging them a higher rate.

"I'm not worried about Nazi eugenics. I've read about that and I'm well aware of what it is. I'm not worried about us making races disappear. What I'm worried about is what are we going to do about insurance companies who insist on this ... companies who advertise that as "Every responsible mother would do this,'" Caplan said.

Once the technology was widespread, there could even be reluctance to provide services such as group homes and therapy for those who were born with disabilities, Caplan said.

"(People would say) "We're not going to pay for them through our taxes,'" he said.

There is also some concern that people should know more about genes and disease and what they're intended for before they go mucking around with them.

Even the worst diseases, like Huntington's disease, may have served a purpose in the past, said Dr. Francis Walker, a professor of neurology at Wake Forest University School of Medicine.

"This gene may actually have served a beneficial purpose at some point," he said.

Huntington's disease affects people in their 30s or 40s. It is a degenerative disease that causes severe damage to the brain. People's personalities change, they become insane and eventually lose the ability to walk and talk before the disease kills them.

In years past, when most people didn't live past 40, the true horror of the disease would not have been realized, Walker said. Instead, the disease would have made people just brave — or crazy — enough to explore new lands, fish in frightening oceans and try other dangerous things. "In the Middle Ages when people didn't live much past 30, Huntington's was kind of a joke. This is really a nondisease up until the last 200 years," Walker said.

The disease has some strange, unexplained effects, Walker said. The gene apparently makes women more fertile, though no one is sure why.

"Diversity in the genome is probably a good thing. In some situations this might not be a bad gene," Walker said. "This gene may actually have served a beneficial purpose at some point."

Inevitable conclusion?

No one knows what the future holds, though most people bet that genetic engineering will be part of it. How much a part is a subject that is debated heavily by ethicists and scientists.

"I think it's very likely we will see eugenics go forward but in a different way than in the past. In the past, eugenics has meant involuntary, forced intervention — what's usually called negative eugenics. And it's aimed at groups, black, the retarded, criminals," said Caplan. "But the future of eugenics, at least the future of genetic engineering, is based on a group of people who choose to do something."

In nearly every case, people are choosing to do something based on the laudable desire to have healthy children, said Lyerly.

"I hope that in part it comes from the natural desire to have a healthy child. I think unfortunately taken to an extreme, that means the healthiest child, the strongest child, and the most well-adjusted child. I guess it's a reassuring desire in moderation, and I obviously counsel my patients to have the healthiest lifestyle they can while they're pregnant," she said.

"Most women will risk life and limb to assure that their babies are healthy. So I hope it's just a desire of women and parents to have the healthiest child they can possible, and a failure to see the broader picture of what they're doing when it comes to the microcosm of their life."

Stirring Up Academia
Researchers hop on scientific breakthroughs to promote their causes

By Danielle Deaver
JOURNAL REPORTER

A civil-rights organization famous for tracking hate groups such as the Ku Klux Klan is now watching several university professors who have suggested that weak parts of the population should be eliminated through a modern version of eugenics.

Although anyone browsing the Internet can find dozens of sites espousing these beliefs, it is especially dangerous when those thoughts come from respected professionals, said Heidi Beirich, a researcher at the Southern Poverty Law Center who tracks the academic movement.

"One thing these academics can do is they provide justification to people for their racist beliefs," she said. "If you have a Ph.D. after your name, you have a lot more clout than Joe Schmo who's talking about how evil the Jews are."

Many of the researchers, including some from respected institutions such as Florida State University, the University of New Orleans and the University of Western Ontario, study race and its connections to intelligence, criminal potential and psychosis. Some then suggest using eugenics to improve the population.

"I basically concluded if we don't have some type of eugenics program, we may have to limit the population," said Edward Miller, a professor of finance at the University of New Orleans who also studies race.

But intelligent parents don't always produce intelligent children, said Frank Wood, a professor of neuropsychology at Wake Forest University School of Medicine.

"People tried to look at it to see if it was inherited. That turned out to be only minorly interesting. Yes, it runs in families, but familial connections, including genetics, have never been the whole of IQ," Wood said.

A movement begins

After falling out of favor in the 1940s, eugenics started interesting researchers again in the 1960s, said Barry Mehler, a professor of history at Ferris State University in Big Rapids, Mich., and the director of the Institute for the Study of Academic Racism.

"In the '60s, we began to see a real trend toward the new eugenics, with a whole new generation for blatantly racist work by people like (Arthur) Jensen and (William) Shockley," Mehler said.

Jensen, a professor of education psychology at the University of California at Berkley, developed intelligence tests that led him to believe that blacks are genetically destined to be less intelligent.

Shockley, a Nobel Prize winner for his work in physics, researched the connection between heredity and intelligence for years and came to the same conclusion as Jensen.

As more research money became available in the last 10 years, the rekindled eugenics movement accelerated and attracted the attention of the Southern Poverty Law Center, Beirich said.

"It's definitely re-emerging. At the early part of the (20th) century, eugenics was a very big topic of research ... in the last 10 years it's become resurgent because you have the funding," Beirich said.

She said that most of the research money is coming from the Pioneer Fund, a nonprofit organization that pays for the study of heredity and race. The fund supported the eugenics movement when it began in the 1930s and now spends millions each year on research that looks at genetic differences between the races.

The movement is also growing because researchers interested in eugenics are getting better organized, Beirich said.

Much of their work is published in American Renaissance and

Mankind Quarterly, journals that are written in scientific language but don't follow such scientific conventions as peer review - the process that gives other researchers a chance to verify research results.

These researchers also move in the same circles, attend the same conferences, review each other's books and exchange correspondence.

"They know each other very well," Beirich said. "They have gotten their act together. They shouldn't be taken lightly because I think we're going to be hearing more from them in the future."

J. Philippe Rushton, a professor of psychology at the University of Western Ontario, is one of the researchers at the top of the watch list at the Southern Poverty Law Center.

Rushton has researched the differences between the races for years. He has compared IQ scores, brain size and fertility rates of blacks and whites. His studies have led him to conclude that blacks as a whole will never measure up to whites, he said.

"Give them the best opportunities. But I think we have to learn to live with the differences. On average, there are going to be fewer geniuses, fewer people in the top professions," Rushton said.

The Institute for the Study of Academic Racism tracks several academics who are researching race differences and nearly always coming up with results that favor whites.

One recommends "phasing out" people of incompetent cultures. Another claimed that Jews used eugenics to increase their intelligence, verbal skills and ability to manipulate and use propaganda.

Glayde Whitney, a Florida State University professor who recently died, wrote the introduction to David Duke's autobiography. He also did studies that tried to link the number of crimes in a city to the size of its black population.

Genetics plays a role

The other reason for academics' renewed interest in eugenics is the Human Genome Project and other genetic research that promises to make eugenic selection easier.

Classical eugenics - the forced sterilization programs that were popular in the last century - didn't work because it was too difficult to track down and sterilize enough people, said Richard Lynn, a professor emeritus at the University of Ulster in Northern Ireland and the author of Eugenics: A Reassessment.

"The idea of eugenics is that we could improve the population namely in intelligence, more character - it would be geared to have people who have strong moral sense of right and wrong, not criminals - and health. People agree these days that these are to some degree inherited," Lynn said.

Experts still disagree about how much genetics influences intelligence and more abstract qualities. But despite that, some academics still want to use IQ to decide who should reproduce. "Initially this idea was that the classical eugenics was to alter people's reproductive pattern without much effect. But the reproductive techniques are much more promising. They can act on selecting the genes," Lynn said.

Nearly all of the researchers in this field point to efforts that they say are under way to select certain fetuses. Pregnant women have amniocentesis to find out if the fetus they are carrying has Down syndrome or other genetic diseases. If fetuses do have the disease, women can choose to knowingly give birth to the child or have an abortion.

"That kind of got going and is seen by some as the thin edge of the wedge, the new eugenics," Lynn said. He thinks that genetic screening is inevitable and will be mandatory at some point.

"I believe it will become possible for women to use in vitro fertilization to grow a number of embryos in glass dishes, of course, and it would be possible to evaluate the genetic specifications of these," Lynn said. "The information would cover those conditions - intelligence, personality, personal health, maybe personal appearance, height, sporting and musical abilities - the genetic potential of these embryos would be printed out and the woman or couple would choose which one to implant."

The ethical questions that would come up - should parents be allowed to have children who are known to have disabilities - would probably be resolved in favor of the unborn child, he said. "You've got a balance between individual freedom. You also have to consider the interests of the unborn child. My own feeling would be to try to stop this from happening from the point of view of the child. Common sense tells us it is better to be hearing than deaf," he said.

That type of thinking is eugenics, and it's a good thing, Lynn said.

"People use the phrase 'back-door eugenics.' They say this biotechnology is eugenics coming in through the back door. No one calls it eugenics, but let's face it, it is eugenics," he said. "A lot of people think this is eugenics and think it is a good idea."

There are others who believe that this will happen. Respected academics such as Lee Silver, a professor of biology at Princeton University, have advocated genetic engineering as a way to create a better human.

According to his writings, Silver thinks that the genetically engineered people would eventually make up about 10 percent of the population. The "Genorich," as Silver calls those people, would control the world. The other 90 percent would work at low-paying jobs at the bottom of the socioeconomic ladder. Silver, who has been criticized for his views, declined to comment for this story.

Political considerations

Some see the new eugenic views as racism backed up by bad science.

"The ironic thing is, most of these people are not geneticists," Beirich said. "Rushton literally spends a lot of his time measuring penis size and head size."

Mehler said he finds the quality of the work particularly offensive because of his own background. He studied with Jerry Hirsch, a behavioral psychologist who spent 30 years breeding fruit flies searching for one very specific characteristic - the direction they fly when placed in a bottle.

"That is science," he said. "They are oversimplifying things. People who knew about science weren't saying nature or nurture because it wasn't that simple."

Observers also object to the conclusions that the researchers draw because often they seem to end up in the same place as before - finding blacks to be inferior.

"The move has been toward a genetic racism. However, what usually happens is the same characters fall into a genetic type. You're talking about the genetic inferior, and they turn out to be the blacks and Hispanics ... the genetic inferior turn out to be the same groups that have always been persecuted," Mehler said.

People have resorted to more than name-calling. The researchers who study these areas and come to these types of conclusions have almost all been discredited at their universities or in their communities. Most stay at their universities over the protest of other professors and community members only because they have tenure. Miller, the University of New Orleans professor, faced a barrage of negative publicity after he wrote a letter to a weekly publication about his views on eugenics. People all over Ontario have called for Rushton to be fired.

"Dr. Miller's opinions on race and intelligence are his, and his alone, and I personally find them to be reprehensible.... There has been no university support for this research," Gregory M. St. L. O'Brien, the chancellor of the University of New Orleans, wrote in a 1996 letter to the student newspaper.

For the researchers, such reaction is often disappointing.

"I'm saddened by it. I think it's completely untrue. I have to be philosophical and accept that it comes with the territory," Rushton said.

"The people don't like the findings. But instead of combating the findings with alternative science, they basically resort to name-calling."

Painless and Permanent
Detractors fear that many women will not be clear about the consequences of cheap new sterilization method

By Kevin Begos
JOURNAL REPORTER

Two North Carolina companies hope that a new method of sterilization will provide safe and cheap birth control for women around the world, but shadows from the era of eugenic sterilization have complicated the discussion.

Quinacrine was originally developed as an antimalarial drug, and it comes in the form of small pellets. When inserted into a woman's uterus, the pellets cause inflammation and then scarring of the fallopian tubes that block the passage of eggs. The cost is about $5 for each treatment.

The U.S. Food and Drug Administration has not approved the method for domestic use, but more than 100,000 women have used quinacrine overseas. Phase One clinical trials are under way here.

Quinacrine is in the testing stage but last month the FDA approved Essure, another sterilization method that doesn't require an operation. Essure is a flexible device that a gynecologist inserts into the fallopian tubes that forms a block. Essure was approved without controversy.

Dr. David Sokal of Family Health International in Durham said that his company hopes to get FDA approval for quinacrine, but that might take 10 years. "If such a method is ever approved by the FDA you'd want to be very careful in the marketing and distribution. And make sure women knew what they were getting into," Sokal said.

Family Health International is a major player in international birth-control programs. The company won a five-year, $87 million contract from the U.S. Agency for International Development in 1999.

But two former employees of the company who believed that quinacrine was ready for widespread human use ran into trouble when a series of media reports said that women in the Third World were being given quinacrine before proper tests had been finished.

Dr. Elton Kessel and Dr. Stephen Mumford, the head of the Center for Research on Population and Security in Research Triangle Park, started to ship quinacrine for birth-control use in the mid-1980s, but the FDA eventually said that they were violating U.S. law by shipping abroad from a storehouse in Mumford's home in Chapel Hill.

India and Chile have banned the use of quinacrine pellets, and the World Health Organization called for further tests before it is used anywhere as a birth-control method.

"The potential for abuse of quinacrine has been grossly overplayed," Mumford told the *Winston-Salem Journal* recently. But though Mumford's and Kessel's Web site contains a wealth of data about the product, it also shows why some critics are uneasy about the rush to distribute the product before full clinical tests are finished.

A paper on the site about the benefits of quinacrine was written by Sarah G. Epstein -the daughter of Dr. Clarence Gamble. Gamble helped found the Human Betterment League of North Carolina in 1947 to promote eugenic sterilization, and Journal research shows a long history of abuses in the N.C. sterilization program - abuses that Gamble consistently glossed over.

Epstein herself makes the link to her father's work, writing that "his battles ... with birth control opponents remind me today of what we are facing with (quinacrine)."

Both sides in the debate over quinacrine are being somewhat dishonest, said Johanna Schoen, an assistant professor of women's history at the University of Iowa.

"Sarah Epstein's rousing endorsement ... ignores the fact that any easy contraceptive method - and particularly one as inexpensive as quinacrine and as easy to administer - carries with it a higher likelihood of abuse," said Schoen, who has done extensive research on birth-control methods.

"The realities of family planning politics around the world are likely to lead to a situation in which especially poor women get pushed towards the use of quinacrine - or even worse, are given quinacrine without their knowledge."

But Schoen pointed out that quinacrine itself isn't good or evil.

"The one problem is that the safeguards that you have to have in order to assure that it isn't abused might have to be higher," she said.

Still, Schoen said that quinacrine supporters are correct in pointing out that there are plenty of poor women who don't have access to cheap and effective family planning methods and that quinacrine might be a godsend for them, as well as for women who want sterilization without surgery. The key, she said, is for any birth-control program in the world to offer women a variety of options, instead of pushing them toward one solution - especially a permanent solution.

Sokal, of Family Health International, agreed with that point. "Absolutely. It's something we've been talking with women's groups to prevent abuse if it (quinacrine) ever does get approved," he said. "With all contraceptive methods you need good options."

Bedfellows in the Cause
Major backer of eugenics program also financed birth control clinics

By Danielle Deaver
JOURNAL REPORTER

Though North Carolina ordered the third-largest number of eugenic sterilizations in the country, the movement that people now scorn may have pushed the state to become first in the fight to give women greater access to birth control.

In 1937, public-health departments in North Carolina became the first to offer birth control to the poor.

"North Carolina was kind of in the vanguard as far as legislation and funding of birth control," said Simone Caron, an associate professor of history at Wake Forest University.

Money for the program came from Dr. Clarence Gamble, an heir to the Procter & Gamble fortune who later helped pay for efforts to expand the state's eugenics program.

Gamble worked with the N.C. Public Health Association to implement the birth control program. During the group's annual meeting in 1938, Dr. G. M. Cooper, a member of the state board of health, introduced Gamble. Cooper also introduced the possibility of using birth control - on willing people only - in a way that would further the cause of eugenics.

"All of you are familiar with the program that we have sponsored at the State Health Department in the Division of Maternity and Infancy Welfare, and in the last 18 months we have undertaken in the bureau to give assistance to those organizations ... in an effort leading toward adopting the European idea of birth control, that is, a positive breeding of better family children, more of them, and the curbing of the breeding of the undesirables," Cooper said.

The birth control offered was foam powder, a method that didn't need to be prescribed or overseen by doctors. The powder was inserted in a sponge and turned into foam inside the vagina.

Because doctors weren't needed, they didn't encourage women to use it. Still, the method caught the attention of women. North Carolina ended up with 56 state-created birth control clinics by the end of 1938 - 13 percent of the clinics in the country, even though the state had only 3 percent of the population, according to an article by James Miller in the Population Research Institute Review.

Gamble stopped paying for the foam birth-control program in the early 1940s. But by that time the state's reputation for being on the cutting edge of birth control had attracted another influential eugenicist, Frederick Osborn.

"The state of North Carolina has been the first to provide contraceptive information through state agencies. That seems to indicate a fearless and imaginative leadership somewhere, probably the chief public-health officer with good backing at the top," Osborn wrote to Dr. William Allan, a professor of medical genetics at the Bowman Gray School of Medicine, in February 1940.

Those qualities made Osborn think that North Carolina might be a good place to set up the first of what he hoped would become a nationwide chain of departments of heredity - county agencies where people's heredity could be charted and decisions made about whether they could reproduce.

"I don't know how long this condition will last. While it does, is there an opportunity to begin your work now with the State Department of Public Health? On a conservative basis, limited to physical defects known to be hereditary ... it would be appealing and could be kept non-controversial," Osborn wrote.

North Carolina was not the only place where the eugenics and birth-control movements became entwined. Margaret Sanger, the founder of Planned Parenthood, got involved with the early eugenicists while trying to raise money for efforts to make birth control more available.

To some, it was her interaction with the eugenicists - who were generally considered to be influential - that allowed the birth control movement to progress.

But some have speculated that she may have shared the views of the eugenicists.

"She was very involved in the eugenic movement. She began more as a feminist. She thought in the early years that birth control was good for a woman's health. That got very little attention," Caron said. "She more or less sells out the feminist cause and becomes a eugenicist."

LEGAL: N.C. can still sterilize retarded adults

By Danielle Deaver
JOURNAL REPORTER

Despite the furor over the eugenic programs run by more than 30 states during the last century, North Carolina still has a law that allows sterilization of mentally retarded adults without their consent.

The law, G.S. 35-36, allows the sterilization of mentally retarded people in state institutions. The parents of a patient or the head of an institution can request sterilization through a petition that must be approved in district court.

Sterilizations can be requested if the operation is "considered in the best interest of the mental, moral or physical improvement of the resident or patient, or for the public good." The statute has been challenged in court several times in the past 25 years, but never successfully.

The state does not keep track of the number of sterilization petitions approved every year, according to officials with the Administrative Office of the Courts. Records are kept only in the courthouse where the petitions are approved.

The N.C. Department of Justice would become involved in any petition that involved a state institution, said John Bason, a spokesman for the department. No one in the office can recall any petitions from state institutions during the past 15 years, he said.

But there have been requests that the court has acted upon, said Ellen Russell, the director of advocacy and chapter services for the Arc of North Carolina. Officials with the Arc, who sometimes consult on the cases, don't approve of involuntary sterilization for any reason, Russell said.

"You go back to the (19)30s or '40s and people were trying to cure society's ills by manipulating the human race. And that is - from a policy standpoint - still what that is," Russell said. "As with any other policy, there may be individuals who need one thing or need another on a case-by-case basis."

The reasons that people request sterilization vary, Russell said. Some parents think that their daughters can't handle the responsibility associated with a monthly menstrual period. Some worry that their daughter will become pregnant through either consensual sex or through sexual abuse.

"Certainly sexual abuse of women with disabilities is a tremendous issue. But sterilization doesn't in any way solve that problem. It solves the parental concern about the result of that abuse," Russell said.

Sterilization, besides being an invasive and potentially dangerous way of solving these problems, also takes away a choice that some disabled people would like to be able to make - whether to have children, Russell said.

Editorials

Against Their Will
December 10, 2002
WINSTON-SALEM JOURNAL

The current Journal series on eugenics and North Carolina's sterilization program opens some old wounds that many might prefer to keep closed. It was not a proud period of state history, and it is frightening to think that it lasted into the 1970s, after many other similar programs had been abandoned for many years.

It is past, and cannot be undone. It would be comforting, perhaps, to ignore this piece of our history and look forward to what a consensus of experts believes is a bright future. But this is the sort of cautionary tale that needs close study because it could all too easily be repeated as science races ahead of our capacity to understand its moral and ethical consequences. Someone said as early as 1670 that "the road to hell is paved with good intentions." The warning still rings true.

There is no doubt that many of the people involved with these programs believed they were working to make the world a better place. The idea was that many human defects were hereditary, and if you could prevent defect carriers from reproducing, you might eventually eliminate the defects.

Since World War II and Hitler, the consequences of such thinking should have given pause to advocates of these programs. They had a very dark side to them. In North Carolina, where before the program was finally ended, some 7,600 individuals had been sterilized, there was too little caution, too little thought of unintended consequences.

In theory, sterilizations were generally consensual, and circumstances of each case were carefully reviewed. In practice, that didn't happen, and in some cases, it was clear that racial bias, a belief that blacks were inferior, motivated and justified the program. We have, one must hope, come a long way from those days. But the journey is not ended. And that is why scrutiny of the past is not voyeuristic or sensational; it is a wake-up call.

It is amazing what medical science can do today that it couldn't 40 years ago. But 40 years from now, what medicine will be able to do is virtually unimaginable. Perhaps the most important question we face is: If we can do it, should we? Put another way, at what point does genetic tinkering jeopardize our basic understanding of what it means to be human?

This is not a matter for scientists to decide. They will provide us with the information that we as a society need to make these decisions. What would it mean for our grandchildren if their parents could decide, with no input from them, what sort of genes would influence their development? Are we tempting God, flying, like Icarus, too close to the sun?

The three big lessons to be learned from this series are:

First, we don't know as much as we think we do about cause and effect. Second, unintended consequences are virtually inevitable. Third, the best-intentioned ideas frequently fail in their implementation, often because their original purpose is distorted by people pursuing other agendas.

It is good that the Wake Forest School of Medicine is not attempting to dismiss or excuse its role in eugenics and sterilization efforts. The dean of the school, William Applegate, has promised to set up a review committee of faculty and administrators to examine the school's role in these programs.

We look at history with the 20/20 vision of hindsight. We think that if we knew then what we know now, things would have been different. We need to remember that this could be said of every generation throughout history. A little humility seems appropriate.

Against Their Will II

December 11, 2002
WINSTON-SALEM JOURNAL

The story of what went wrong with the Human Betterment League, founded in 1947, can be viewed as a classic example of why democracy works better than any other political system yet devised. Where the public weal is concerned, decision-making should be participatory and inclusive. It wasn't in this instance, and the result was an awful mistake.

The league, founded by James G. Hanes of Hanes Hosiery in Winston-Salem and others, promoted eugenics - improving the human race by controlling hereditary factors. The method of choice was sterilization of the "feeble-minded" and "promiscuous."

Eugenics, as the Journal series this week details, was in those days more theory than science.

This was not Winston-Salem's finest hour. The league pursued its objectives with skill and determination, and sterilizations rose dramatically in North Carolina, even as they were declining everywhere else. In 1945, there were 117 sterilizations. In 1954, they peaked at 700, while the science used to justify the programs was being refuted.

Heredity was a lot more complicated than the advocates of such programs understood. More seriously, the programs, designed to be voluntary but too often mandated in practice, violated human rights and basic humanity.

Yet the league and its objectives drew strong support from this and several other major newspapers around the state. The net effect of having decisions made in this community by a handful of leaders, one of whom, at the time, owned the newspaper, was to limit public discussion of the issues. Maybe, just maybe, a more public, democratic approach would have resulted in voices of caution being heard.

One suspects a pretty wide gap of experience between the people subjected to the sterilization program and its promoters. While many of those promoters appear to have been people of good faith and even people determined to give a hand to those in need, the help provided was rarely hands-on.

The promotion of the program by high-profile citizens may also have given the bureaucracy that ran the program a false sense of righteous power. Clearly, there was a "we know what's best for you" attitude in some places. Only in recent years has the delivery of social services become infused with the effort to involve the client in decisions about what services are needed.

It would be hard to argue that Winston-Salem and its residents were not net beneficiaries of the patriarchy that ran things for so long. Because the leaders did much good, others probably questioned their wisdom less as time went on. And one senses a provincialism, an isolation from the rest of the world, in the way that the Human Betterment League developed in the face of growing doubt about the science, never mind the moral authority, behind the concept.

We are, being human, subject to erring. The best and brightest among us make mistakes. If ever the oft-maligned skepticism that journalists embrace to keep them asking questions was ever justified, this was an instance, and journalism came up short. In a sense, then, the series is an effort, belated as it may be, to make amends.

It is unfortunate that power, even power wielded to do good, tends to breed arrogance more often than humility. Few of us know as much as we think we do, and it is therefore better that we include more points of view, especially when we make life-changing decisions.

Against Their Will III

December 12, 2002
WINSTON-SALEM JOURNAL

The faces, the voices, the stories. This newspaper's series about North Carolina's eugenics sterilization, whose last installment is running today, is disturbing in many ways and for many reasons, but it's the human elements it portrays that keep coming to mind.

What happened in North Carolina, and especially in Winston-Salem and Forsyth County, is a cautionary tale with lessons that ought to be heeded as we try to figure out the best ways to take advantage of today's cutting-edge science. It's a reminder of how easily good people - community leaders, journalists, doctors, politicians and ordinary citizens - can tolerate and even promote something that's terribly wrong.

The revelations of how North Carolina sterilized more than 7,600 people, often against their will and on the flimsiest of pretexts, should provoke serious soul-searching and discussion. What does this history say about our state, our community, our heritage? Could something like this happen here, today?

The eugenics movement and its failed attempt at social engineering have been discredited. But how does our society today deal with the mentally ill, the developmentally disabled, the misfits, the poor, the elderly, the neglected and abused children? Do we treat those who are different from us because of race, ethnicity or sexual orientation as we should, or do we still harbor prejudices? Are we paying as much attention as we should to the decisions that are being made about such issues as cloning, genetic engineering, stem-cell research - or do we just leave all that to somebody else?

Such soul-searching and discussion are important, and the need for such activities is high on the list of reasons why it would be wrong simply to let the story of eugenics and forced sterilizations in North Carolina remain cloaked in secrecy.

But while so engaged, we should not let those faces and voices fade from our memories. Making up the big picture of what happened in North Carolina are a lot of individual people who were unfairly denied one of the most basic of human rights, the right to have children.

One way to honor the victims who are dead is to try to avoid similar mistakes. But because North Carolina continued its sterilization program well into the 1970s, even when such programs were being discredited in many other places, many of its victims are alive today. What can we do for them?

In May of this year, the governor of Virginia, which had the nation's second highest number of forced sterilizations, apologized for his state's role in the "shameful effort" and unveiled a memorial to the 18-year-old unwed mother who was the first person sterilized under Virginia's 1924 law. Earlier this month, the governor of Oregon apologized for that state's forced sterilization program and established Dec. 10 as Human Rights Day there. Here in Winston-Salem, as the story has unfolded, the Journal's publisher has apologized this week for this newspaper's role in legitimizing the eugenics sterilization program.

Some may dismiss such public apologies and memorials as shallow gestures. They are not. They serve the purpose of forcing us to confront our history, with the hope that we will learn the lessons it offers. They also tell the living victims that society does value their worth and dignity, that we acknowledge and regret the great wrong that was done to them. That would seem to be the least that the state of North Carolina could do.

Epilogue

Easley apologizes to sterilization victims
December 13, 2002

By Kevin Begos, Danielle Deaver and John Railey
JOURNAL REPORTERS

Gov. Mike Easley apologized last night for the state's role in sterilizing more than 7,600 people through a eugenics program that lasted from 1929 to 1974.

"On behalf of the state I deeply apologize to the victims and their families for this past injustice, and for the pain and suffering they had to endure over the years," Easley said in a statement to the *Winston-Salem Journal*.

"This is a sad and regrettable chapter in the state's history, and it must be one that is never repeated again," he said.

The apology came in response to a *Journal* investigation into the Eugenics Board of North Carolina.

The five-part series, "Against Their Will," concluded yesterday.

The Wake Forest University School of Medicine also has formed a committee to investigate the school's role in the state eugenics movement in response to queries from the newspaper.

The eugenics movement made exaggerated claims that mental illness, genetic defects and social ills could be eliminated by sterilization. Children as young as 10 were sterilized under the state program, which was often characterized by coercion and flawed intelligence testing. By the 1960s, it mainly was targeting young black women. The North Carolina program was the third largest in the country, after California and Virginia.

Nial Cox Ramirez, 56, of Riverdale, Ga., who was sterilized at age 18 in 1965, thanked Easley for the apology.

"That is good, that is very good. I appreciate that, I really do. That makes me real happy. What a long time. A long time coming, but it came," Ramirez said.

Bertha Dale Midgett Hymes, 52, sterilized when she was 17, said she forgives the state, "but I don't want them to do anybody else like that."

Elaine Riddick Jessie, 48, of Atlanta, who was sterilized at age 14 in 1968, said that the apology was "fantastic," but still had her doubts.

"They are embarrassed - I mean, they should be. I'm glad that they apologized, but do they really mean it?" Jessie asked.

Carmen Hooker Odom, the secretary of the N.C. Department of Health and Human Services, joined Easley in offering an apology.

"As secretary, I want to be very forthright in issuing an apology on behalf of the department to all of the victims and their families," Hooker said. The eugenics board operated under previous agencies whose duties are now covered by the Health and Human Services Department.

Skip Alston of Greensboro, the president of the North Carolina branch of the National Association for the Advancement of Colored People, said that an apology might not be enough.

"The state, if they allowed this to happen, they should do more than apologize," Alston said. "They should seek other remedies to correct that wrong in as much as possible ... reparations and whatever else might be deemed necessary or satisfactory to the victims."

Hooker Odom said: "On the face of it, it (reparations) would be something that I think the state should consider, but since I'm not a lawyer ... I don't know what that would entail. But these people were harmed irreparably through a policy that may not have been endorsed by the governors of that time or the state agencies. My heart says that these people were greatly harmed."

Earlier this year, Virginia became the first state in the nation to apologize for its sterilization program, and Oregon followed suit last week.

The Journal gained access to thousands of records from the eugenics board, but many more are still under seal in the state archives. Until the newspaper's investigation, few details had been known about the sterilization program. More than 2,000 people age 18 and under were sterilized in many questionable cases, including a 10-year-old who was castrated. The Journal has made a formal request for access to all the records.

Hooker Odom said that she does not think any further investigation is necessary, and the records that detail the program could be opened only if the privacy of the victims can be guaranteed.

Yesterday, several religious and secular leaders had joined victims in asking for an apology.

The Rev. Jerry Pereira of Swannanoa, the president of the state's largest group of Christians, the 1.2 million-member Baptist State Convention of North Carolina, condemned the sterilization program.

"That policy was morally bankrupt, ethically flawed.... It's just terrible," Pereira said. He emphasized that he was speaking personally and not for the convention.

"Unless we know history, we cannot in the future avoid errors that were made," said Jack Fleer, a professor of political science at Wake Forest University.

"In this particular instance it appears errors were made," Fleer said. "And I think somebody has to explain that or justify it."

N.C. first to weigh eugenics amends
State committee will consider reparations for those sterilized
February 11, 2003

By Kevin Begos, Danielle Deaver and John Railey
JOURNAL REPORTERS

RALEIGH-North Carolina has become the first state to officially consider paying reparations to victims of a eugenic sterilization program.

Gov. Mike Easley has appointed a committee to examine the subject, to be headed by Carmen Hooker Odom, the secretary of the N.C. Department of Health and Human Services.

"We need to emphasize the fact that we consider this to have been a dark, dark time in the history of North Carolina and to make sure that people who suffered from those actions are aware of that. Whether there is anything else the state can do to meet the needs of those individuals is something we will look at," Hooker Odom said yesterday.

Hooker Odom said that the committee will look at what other states have done in response to similar sterilization programs, and at information about North Carolina's program before deciding what to recommend to the governor. The committee will include Hooker Odom, her department's general counsel, a representative from the Division of Mental Health and two representatives from the N.C. Department of Cultural Resources, which is responsible for the state archives that house most of the information about the eugenics program.

The first meeting of the committee will be this week.

"I applaud the governor for taking that step," said Skip Alston of Greensboro, the president of the North Carolina Branch of the National Association for the Advancement of Colored People. "Sterilization is the

No. 1 priority on the North Carolina NAACP's agenda, and the governor knows that."

Easley apologized for the sterilizations last month in response to a series of stories in the *Winston-Salem Journal* that provided details about North Carolina's program for the first time.

About 65,000 sterilizations were carried out nationwide as part of the eugenics movement, which claimed that mental illness, genetic defects and social ills could be eliminated by sterilization. In North Carolina, children as young as 10 were sterilized under a state program often characterized by coercion and flawed intelligence testing.

By the 1960s, the program was mainly targeting young black women. The North Carolina program sterilized more than 7,600 people between 1929 and 1974 and was the third largest in the country, after California and Virginia.

Victims of the program were grateful to learn that the state is taking more action.

Bertha Dale Midgett Hymes, 52, of Dare County was sterilized when she was 17. Upon hearing about the committee, she said, "Thank the Lord for that."

Nial Cox Ramirez, 56, of Riverdale, Ga., was sterilized in 1965 when she was 18. Ramirez filed a lawsuit against the state's sterilization program in 1973, but it was later dismissed on technical grounds. "I feel real good about this, because somebody needs to do something," she said.

Legal experts said that the idea of reparations for sterilization victims makes sense.

"My initial reaction is I think it's a very positive step," said Arnold Loewy, a professor at the University of North Carolina School of Law. "It's an effort by the state to undo the wrong that was done, and to the specific people the wrong was done to. I feel the state ought to pay for the wrongs it does to people, and this one's pretty high on the list."

The issue of reparations for slavery is more debatable, Loewy said, in part because the people who were directly wronged under slavery are no longer alive.

Alston said he has talked to Rep. Earl Jones, D-Guilford, and Rep. Larry Womble, D-Forsyth, about drafting a bill asking for reparations.

Womble said he and Jones may co-sponsor the bill. "I don't know how far it's going to go, but I'm going to be drafting legislation.... These people need to be compensated in some way," Womble said.

Though there have been no successful lawsuits over eugenic sterilization in the United States, the provincial government in Alberta, Canada, has paid out more than $142 million (Canadian) to about a thousand victims of its sterilization program. The settlements in Alberta came after a woman won a lawsuit in 1996 and was awarded $740,000 for wrongful confinement and wrongful sterilization.

Paul Lombardo, the director of the program in law and medicine at the University of Virginia Center for Biomedical Ethics, who has written extensively about eugenic sterilization, said he was surprised and encouraged by North Carolina's action.

Virginia was the first state to issue an apology for a eugenics program early last year, followed by Oregon and North Carolina in December. South Carolina followed suit last month. In other places "there have been concerns voiced about opening the state to legal liability," Lombardo said. "I know of no state that has offered compensation, or set up a commission to look into it. Sweden, Germany, and Alberta have, and British Columbia just had a brand new lawsuit."

Lombardo said that more than a dozen states have laws to pay people who have been wrongfully imprisoned. "So it wouldn't be strange that we admit to a mistake and attempt to find some way of addressing the compensation," he said.

HIGH HOPES: Birth-control clinics opened to fanfare in 1938

February 16, 2003

By Danielle Deaver
JOURNAL REPORTER

For nearly 10 years, the letters with the fashionable letterhead from the most stylish of addresses - Madison Avenue in New York City - arrived on the desk of George Lawrence, the head of the N.C. Maternal Health League.

The letters were filled with plans to ignite a birth-control revolution in the state, and with triumph when those plans succeeded.

"We were delighted to hear of the new developments in North Carolina which will mean that your birth control work will go forward quickly and with such splendid backing," read a letter from a member of the American Birth Control League to the head of the N.C. Maternal League. "What a perfect place to have a birth control information given out - in connection with prenatal clinics! You and your committee are indeed to be congratulated."

The first state-sponsored birth-control clinics in the country opened here in 1938. They were lauded as proof of the state's progressive policies, as was the decision to make abortion legal here in 1968, five years before the U.S. Supreme Court legalized abortion for the country.

But in between the two movements, another one existed, based on the idea that the state - instead of the individual - was better suited to make reproductive choices.

The eugenics movement helped put the decision about who should have children into the hands of the five members of the Eugenics Board of North Carolina. About 7,600 people were sterilized in the state from 1929 through 1974 - many of them poor black women who were pressured into making such a decision.

But before the eugenic sterilization program took off in the late 1940s and '50s, North Carolina was admired for its birth-control movement, which was praised in national magazines and used as a model for other states.

"No spot in North Carolina is more than 50 miles from a state-sponsored birth-control clinic. First to promote birth control officially, the state is going at the job in earnest," said an article in The Atlantic Monthly in October 1939.

But that movement ran counter to the ideas that the eugenics movement was founded on. The birth-control movement began in part because of the maternal health problems in North Carolina in the early part of the 20th century.

The state had one of the worst infant-mortality rates in the country. About 66 babies died out of every 1,000 born - well above the national average of 54 deaths for every 1,000 births. Mothers were also at high risk.

In 1930, the state had a population of about 3.5 million people.

There were 832 pregnancy-related deaths for every 100,000 live births in North Carolina in 1930. Currently the rate is about 12 deaths for every 100,000 live births.

Officials blamed the high death rates on such basic problems as bad hygiene, poverty and lack of prenatal care.

People hoped that birth control could combat some of these problems.

"To improve conditions so that dirt, poverty and disease will disappear is the ideal solution. But that will take time. Birth control offers immediate help," read The Atlantic Monthly article.

The women who asked for help were usually those who had given birth every year for several years and had no other way of preventing more pregnancies.

"I am married and 22 years old. Have been married 3 yrs. and have 2 babies. We love our babies very much, but we feel we are not able to care for any more. My husband makes a living salary but that is all. I had never heard of birth control clinics before. You don't know how relieved my mind will be when I learn the sweetest secret of married life and my dear friend it will mean the happiness of my home I'm always afraid and it makes my husband ill and cross," wrote Mrs. C.J. Blackman in an undated letter.

But after 1940, little was written or said about the clinics. Instead, the state focused on its eugenic sterilization program.

It is not clear from the documents preserved in the state archives why the birth-control movement received so little attention while the eugenics movement took over. But even at its beginning, the state birth-control movement struggled with too many demands and not enough money.

"Following the announcement of the availability of the consultant services, the difficulty has been to meet the requests from local health officers," said an article in the American Journal of Public Health in October 1938.

The biennial reports from the Eugenics Board of North Carolina and its handbook will be available online beginning Monday at the State Library Web site: http://statelibrary.dcr.state.nc.us

Some eugenics patients died after surgery
February 16, 2003

By Kevin Begos
WINSTON-SALEM JOURNAL

North Carolina's eugenic sterilization program experimented on patients, damaged lives and it ended some, too, according to documents released to the *Winston-Salem Journal* last week.

In the 1930s a "single white female, age 21 years, died from post-operative intestinal obstruction ten days after" her sterilization operation.

Another 30-year-old married woman died two days after her operation, and a 17-year-old girl died from "locked bowels" (accompanied by gangrene) five months afterward.

Gov. Mike Easley set up a committee last week to investigate the eugenics program and consider reparations or counseling services for victims. The program sterilized more than 7,600 people from 1929 through 1974. Many operations were done against the patients' wishes and some were performed on children as young as 10. For years painful memories of the program haunted whole families, another document shows.

A 1951 entry from minutes of the Eugenics Board of North Carolina concerns a sterilization petition for a 14-year-old girl whose mother and grandmother had also been targeted by the program.

"The maternal grandmother was feebleminded. Permission was granted for her sterilization. She died under the operation. Arrangements were made to sterilize (the) mother, but after the death of the grandmother, the operation was dropped."

Other victims struggled to cope with experimental operations.

"One white female was sterilized by X-ray and one by radium implantation," and the woman given the implant complained in a follow-up, "Now suffer from high blood pressure and hot flashes." The woman was "not satisfied" with the outcome of the sterilization, the follow-up says.

The disclosures about the North Carolina program come from a study conducted in 1940 by Eleanor Welborn, a graduate student at the University of North Carolina at Chapel Hill, and from more than 1,400 pages of documents that were released by the N.C. State Archives. With the cooperation of the eugenics board, Welborn did a follow-up survey of 183 non-institutional sterilization cases in North Carolina. She wrote that "three patients ... died from intestinal obstruction shortly after the operation, and a number of patients stated that they had certain health or sex problems which they believed to be the direct result" of the operation. The official cause of death for the woman who died two days after her operation was listed as pneumonia. Welborn wrote that "there is a real need for follow-up physical examinations of sterilized persons at regular intervals ... to aid the patient in combating any effects of sterilization which may damage their health or even endanger their lives."

Carmen Hooker Odom, the secretary of the N.C. Department of Health and Human Services, is head of the Eugenics Study Committee, which met for the first time Thursday. It could make recommendations to Easley as early as May or June. "This just gives another reason why we should be dedicated to making sure we figure out what is fair and just for the victims," Hooker Odom said about the deaths and the experiments. Paul Lombardo, the director of the program in law and medicine at the University of Virginia Center for Biomedical Ethics, said that abdominal surgeries - including sterilization operations - were much riskier before a wide range of antibiotics came into use after World War II.

But Welborn's study was based on a review of just 183 non-institutional sterilization operations. If three deaths occurred in that small group, that would be an extremely high rate, Lombardo said. In Virginia, records indicate a death rate of about 1 for every 1,000 sterilization operations before World War II. Johanna Schoen, a professor at the University of Iowa who has been researching the North Carolina eugenics program for 12 years, said that the quality of medical care may have been below average in the sterilization program.

"It is probably safe to assume that the people who were sterilized did not have the best surgery available," she said. Welborn's study said that the most frequent cost for female sterilizations in the 1930s was $28 - about $360 today. Schoen said that probably meant that inexperienced surgeons performed many operations. Lombardo, who has written extensively about eugenic sterilization, said that more information is needed to put the experimental surgery on the two women in context.

"So much of what went on then and goes on now (in medicine) is a trial-and-error kind of thing. It doesn't necessarily have to be malignantly motivated," he said of experimental surgery. "It usually depends on the context of what the doctors were trying to do."

Another test of intent is to look at whether there was any attempt to hide what went on, Lombardo said. Some of the deaths were obliquely mentioned in the 1938-40 biennial report of the eugenics board, but the experimental operations were not.

Efforts to investigate the program - which was often marked by coercion and faulty intelligence tests - started in earnest after a series of stories in the Journal in December that exposed previously unreleased details. Easley apologized for the program as a result of the series.

More than 100,000 pages of documents related to the eugenics program are stored at the state archives - many under "lock and key" because of concerns about medical privacy. The archives have made some records public, but ruled that medical reports from the sterilizations cannot be reviewed by anyone. Hooker Odom said that the Eugenics Study Committee may take a harder look at the confidentiality issue because of the deaths and the experimental operations.

Lombardo said that confidentiality laws are meant to protect the patient, not to hide medical issues. If a patient's name, age, race, residence and all other "identifiers" are stripped from records, it would not breach any ethical guideline, he said.

"If the information has been de-identified, I'm not sure what argument you make on privacy grounds that would prevent you from releasing it," he said.

Virginia was the first state to issue an apology for a eugenics program, early last year, followed by Oregon and North Carolina in December. South Carolina followed suit last month. In other places "there have been concerns voiced about opening the state to legal liability," Lombardo said. "I know of no state that has offered compensation, or set up a commission to look into it.

"I think it's extraordinary that (Easley) has taken the step he has, and is following up in good faith," Lombardo said. He cautioned that for any investigation or reparations program to work, it will have to stay focused.

Some of those ordered sterilized in North Carolina wanted the sterilization operations, he said, and the focus should be on people who were sterilized against their will.

"I think for practical reasons we should stay away from other kinds of damages - including deaths," Lombardo. "(Reparations) should be for people who are alive and not some of their relatives, or relatives of people who died."

Easley asked the Eugenics Study Committee to investigate how the program unfolded in North Carolina, how to prevent a similar occurrence in the future and how to redress the wrongs. Financial reparations are a possibility, but some supporters question whether it's the right time to debate financial compensation given the budget crisis. The state is facing a deficit estimated to be as much as $2 billion over the next two years.

Nationwide about 65,000 sterilizations were performed as part of the eugenics movement, which sought to eradicate mental illness, genetic defects and such social ills as out-of-wedlock births. The North Carolina program was the third-largest in the country, behind California and Virginia.

Little Notice and Less Explanation

The state's eugenics board ordered sterilizations in its last years even as some members sought reforms
February 16, 2003

By John Railey, Kevin Begos and Danielle Deaver
JOURNAL REPORTERS

The state eugenic sterilization program spent its last years in a private struggle to change its course away from a pattern of operations ordered without proper consent, ones performed on children and ones based on flawed IQ testing.

More than 1,400 documents released last week to the *Winston-Salem Journal* tell for the first time the story of the end of the Eugenics Board of North Carolina, which ordered the sterilizations of more than 7,600 people, often young women living in poverty, from 1929 through 1974. It continued to order the operations through its last years, but reforms advocated by some board members might have spared Elaine Riddick Jessie of Atlanta, who was sterilized in Chowan County in 1968 when she was 14, as well as hundreds of others. "It was just too late," said Jessie, who was sterilized shortly after the birth of her first and only child. "The damage had been done."

Almost 35 years later, Jessie finally has the chance to talk to state officials about what happened to her. A committee appointed by Gov. Mike Easley to consider reparations and other methods of compensation for those sterilized wants to hear from her and as many of the others as it can. The Eugenics Study Committee held its first meeting Thursday.

Easley apologized for the program in December in response to a Journal investigation, and widespread news coverage has followed.

But it attracted little attention in 1974 as the N.C. legislature moved to end the program. The law ending it didn't take effect until 1975. Even in the last year there was no public acknowledgement from Gov. Jim Holshouser and no explanation from the board to those it had sterilized in a relatively routine fashion, with little direction but to sterilize as many as possible - until its last years.

"These things that we do come in on kitten's feet," said Wade Smith of Raleigh, a former state representative who worked on a bill that was instrumental in disbanding the board. "They come in and we're hardly aware that they're going on. We should be aware that they're going on."

Some who served on the five-member eugenics board were happy it was over.

"I just know I was glad to be off that thing," said Bonnie Allred of Raleigh, who represented the state director of social services on the board during its last days. "I just didn't want to participate in something like that. It was sort of like playing God."

Turning the tide

Allred is one of the few former members who will talk publicly about the board's work. Of the former members still alive, others say they remember little of it, or that they don't want to talk about it.

Allred, like many who served on the board, attended only a few meetings, acting on orders from her boss. Seasoned agency heads also served.

The board was part of the Department of Public Welfare, and its members were not appointed by governors. Instead, state law set the membership at one representative each from the attorney general's office, Dorothea Dix Hospital, the state Department of Public Health, the state Department of Mental Health and the state Department of Public Welfare (later the state Department of Social Services).

The board met monthly in Raleigh to consider petitions for sterilizations from social workers statewide, rapidly reading brief case descriptions and usually voting to sterilize. Dedication on the part of the members was sometimes lacking, according to one of the documents released by the N.C. State Archives in Raleigh.

"The members are all busy with their own work," Ellen Winston, a board chairwoman, wrote in a 1955 memo. "They, therefore, have little opportunity to give thought to their responsibility to this other important program."

In other documents, board members speak of their dedication to a program inspired by the eugenics movement, which made exaggerated claims that mental illness, genetic defects and social ills could be eliminated by sterilization. Some members occasionally voiced opposition to the program, but that was rare until 1970. The new decade dawned with a growing awareness of civil rights, particularly the rights of women and blacks - who by then made up the majority of those sterilized. There were other changes as well.

To the public, eugenics board members continued to say only positive things about their work. The documents released last week tell a different story.

"With the liberalization of sterilization and abortion laws and recent developments in contraceptive measures and their increasing availability, we feel that many cases coming before the board could be handled in the community through existing laws and resources," Clifton Craig, a board chairman, wrote in a 1970 letter.

By the next year, he was proposing something more radical. "It continues to be my belief that the Eugenics Board is outmoded in law and operation and that new laws must be adopted," he wrote in a letter to his fellow board members. Craig, who was also the head of the state department of social services, wrote that several organizations "are expressing concern over the present Eugenics Board and the laws relating thereto."

Among those groups, he wrote, were the N.C. Medical Society and the Human Betterment League - an organization based in Winston-Salem that had been created to promote the sterilization program and spent its last years touting birth control.

Almost every other state had ended or sharply cut back their sterilization programs right after World War II. Even in their attempts to reform North Carolina's program, Craig and some of his fellow board members were decades out of step with science and social policy.

Yet rather than resign his seat, Craig stayed on the board, trying to change it from within.

Under his leadership, the board continued to order sterilizations, but at a slower rate. At a June 1972 meeting, it ordered five operations and turned down petitions for two others. But at the same meeting, it adopted several policies that would limit the sterilizations it had once ordered freely.

"Alternatives to surgical sterilization such as oral contraceptives and intra-uterine devices should be considered and reasons why such measures are inadequate or inappropriate should be noted in the material presented to the Eugenics Board with a petition for sterilization," according to minutes from that meeting. In a sense, the board was returning to its roots. In the 1930s, members had a flurry of correspondence with a New York agency that promoted birth control. Yet the board never embraced that idea, focusing instead on sterilization.

At the June meeting, the board also mandated that "approval for sterilization will be granted only rarely when the patient or family opposes the procedure" and that social workers produce more documentation to support their petitions for sterilization. "There must be actual indications that conception is a likely possibility," it said.

The board ruled that "IQ alone is at times a questionable indication of mental retardation. A request for sterilization based on IQ alone, especially when the figure is over 55, indicates the need for further evaluation."

The policy changes made concrete a direction that the board had been taking since 1970. Since then, it had been routinely turning down petitions for reasons of questions on IQ, other options and lack of documentation, as well as because those targeted for sterilization were too young. And although the board had long relied on summaries that condensed the lives of those to be sterilized down to a page, it now asked for "photocopies of all material presented with petitions." Those decisions came too late for Jessie and others like her. Just a few years earlier, Jessie had been sterilized after board members read a few paragraphs that included lines about "reports of her running around" and a listing of her IQ as 75. She never got a chance to be heard by the board and her father, a shell-shocked alcoholic, signed the consent form for her operation.

A slower pace

The policy changes caused dissent among many social workers who had grown accustomed to having most of their petitions approved. "I have already felt the rumblings of dissatisfaction from all directions regarding the number of negative decisions being made by the Board," June Stallings, an executive secretary for the board, wrote in a 1971 memo.

Incoming petitions slowed, and so did the approval rate of those petitions. In 1971, the board considered 165 petitions and approved 106. In 1973, it looked at 47 petitions and approved 19, according to board documents.

A July 1973 meeting in which two sterilizations were approved and five turned down was typical. "Due to the patient's youth and the absence of positive evidence that sexual involvement exists, temporary contraceptives were recommended," read one of the rejected petitions.

At a meeting in September 1973, all four petitions presented were either delayed or denied, including one recommending a castration. "It was brought out that recent behavioral research indicates that castration does not necessarily inhibit the sex drive," according to minutes from the meeting.

And at a December meeting, a petition was denied because of "research findings that a high percentage of individuals afflicted with mongolism are naturally sterile. They (board members) believe the chances of Miss -- becoming pregnant are very slim and that surgical sterilization is not justified."

But the board continued to approve petitions for women with mental illness or extremely low IQs who were at risk of getting pregnant. In doing so, it held to a program that had been abandoned decades earlier in most other states.

Critics who took on the eugenics board weren't interested in its efforts at reform, if they even knew about them. Nial Cox Ramirez, who had been sterilized in Washington County in 1965, filed the first federal lawsuit against the board, seeking $1 million in damages. Jessie also filed suit against the board. Both suits would fail, but they led to some public scrutiny of the state's eugenics program. And in 1973, Caspar Weinberger - the secretary of health, education and welfare under President Richard Nixon - issued a memo to state officials that "directed a review of guidelines in order to protect rights of individuals in cases of sterilization."

Here, eugenics board members talked to the N.C. Attorney General's office about revising the state law on sterilization.

Yet ultimately, the end of the board came from outside. Action by the legislature abolished the board and gave district court judges the power to order sterilizations.

Stallings, the board's executive officer, praised the action in an April 1974 memo to board members: "Perhaps you have already heard the good news that 'our' legislation was enacted."

Renee Hill, the chairwoman of the board (by then called the eugenics commission), also approved. "The Commission has been actively involved since 1970 in efforts to bring revisions to these laws and is quite pleased with the outcome," Hill wrote in a memo.

Smith, the former state representative, said that he couldn't remember any members pushing to end their board, but it's likely that was the case.

"I don't get an awful lot of credit," said Smith, now a prominent trial lawyer in Raleigh. "I get some. But I think there must have been lots and lots of people working on this, providing the leadership, and I would just be one of those people who were involved."

One can assume, he said, that there were courageous people on the board doing their best to change it. "I'm assuming that if they stepped down completely, there were certainly people out there who would have believed in this, who would have occupied those seats on the board."

It's hard to know what to do in such a situation, he said. "I think the easier thing to do would be to resign and say, 'I'll have no part of this.'"

Jessie would have preferred that course of action. "They should have just walked away," she said. "It just was something they should not have done, period."

'I want to know why'

By the eugenics board's last year, Craig had left, having retired from the department of social services. He died in 1986.

Allred, who went to a few board meetings in 1974 before becoming a member, said she had never even heard of the program until then.

The few meetings she went to made her uncomfortable, Allred said. "In one sense I felt like it was the right thing to do. In another sense it made you feel inner-conflicted.

"Had it continued, I would not have stayed on the board. I was just not comfortable making those decisions for people."

By then, the pace of the board had slowed even more. It didn't meet at all in January or February 1974 because it had received no petitions for sterilization, letters to members state.

Allred was at one of the last meetings of the board in October 1974. Other members shared her relief that the program was ending, she said.

"My best memory is they seemed to behave as if they were glad it was over. It's just not a pleasant thing to serve on."

That day, the board ordered sterilizations on two more people. They were among the last of generations sterilized through the program. The two ordered sterilized vanished into history, their names withheld for reasons of medical confidentiality. Almost 30 years later, they, just like Jessie, could use the Eugenics Study Committee as a forum to talk about what happened to them.

Jessie looks forward to telling her story. "I want to know why. I want to know if they (the eugenics board members) would have done it to themselves and some of their family members."

Some caution against ban on involuntary sterilization

Repeal of the existing law could have
unintended consequences, they say
February 19, 2003

By Dana Damico
JOURNAL RALEIGH BUREAU

RALEIGH – Advocates for the disabled told legislators yesterday that they oppose forced sterilizations of mentally ill and mentally retarded people when the operations are ordered to prevent the patient from having a baby or making someone else pregnant.

But they cautioned that efforts to strike an existing law that allows District Court judges to order involuntary sterilizations could have unintended consequences.

A guardian for a mentally disabled woman battling ovarian cancer could be barred from letting the woman get a hysterectomy because one effect of the operation would be to leave her infertile, Deborah Greenblatt of Carolina Legal Assistance Inc. told a legislative committee.

Greenblatt's comments came during a hearing yesterday before the N.C. House's Health Committee on a bill backed by Rep. Larry Womble, D-Forsyth. Legislators did not vote on the bill.

Womble said he wants to repeal the involuntary sterilization law because it is a "distasteful" vestige of the state's eugenic-sterilization program. The eugenics movement wanted to eradicate mental illness, genetic defects and social ills, including out-of-wedlock births.

About 65,000 sterilizations were performed nationwide as part of the movement, including more than 7,600 in North Carolina. Only California and Virginia performed more sterilizations.

The Eugenics Board of North Carolina - which ordered the operations for epilepsy, mental disease and "feeblemindedness" - operated from 1929 to 1974. After legislators disbanded the group, they shifted responsibility for approving sterilizations to the judicial system.

The existing law allows involuntary sterilizations for the "public good" or for a patient's "mental, moral or physical improvement."

Parents or guardians of a mentally challenged person or the head of a state medical institution can request the operation.

The law is seldom used. Three petitions were granted in 2001-02. Over the past five years, only one patient at a state mental hospital was sterilized. Dave Richard, the executive director of The Arc of North Carolina, said that the law clearly looks "horrible."

"We don't like the way it is," he said. "We know that people with mental retardation can be parents. They can be very good parents."

But Richard agrees with Greenblatt that any changes should allow for medically necessary operations and should clearly define what those are.

Greenblatt also proposed a penalty for doctors and hospitals that perform involuntary sterilizations without following proper legal procedure.

Advocates say they know of several instances in which health officials accepted a guardian's written consent to sterilize a patient - without judicial approval - because the doctor didn't know the law. They want to make sure that caregivers don't skirt proper procedures in the future.

Carmen Hooker Odom, the secretary of the N.C. Department of Health and Human Services, acknowledged the concerns but reminded the committee of the context behind Womble's bill. Womble proposed the legislation after a series of stories in the *Winston-Salem Journal* revealed details about the eugenics board.

People were sterilized based on social class, race and imperfections, Hooker Odom said.

The legislation was designed "to make sure that that dark history in North Carolina will never, ever happen again," she said. "Please, as you work through this bill ... don't ever lose sight of why the bill was filed in the first place."

Hooker Odom was charged by Gov. Mike Easley to lead a committee to study the state's eugenics program and possible reparations for its victims. The group is scheduled to meet again March 14.

Law that lets judges order sterilizations facing repeal
Womble says 1975 measure 'atrocious, ungodly'
February 19, 2003

By Dana Damico
JOURNAL RALEIGH BUREAU

RALEIGH A rarely used state law that allows District-Court judges to order sterilizations for mentally retarded or mentally ill people without their consent could be wiped off the books under a proposal filed in the N.C. House yesterday.

Rep. Larry Womble, D-Forsyth, wants to repeal a law that he called "atrocious and ungodly."

"It borders on genocide," Womble said. "It borders on communist, on Third World countries, on countries that have dictators that they can do people like that."

The law allows sterilization in cases of a person's "mental, moral or physical improvement," or for the "public good."

It took effect in 1975 after legislators dissolved the Eugenics Board of North Carolina. The board - which operated with little oversight - authorized more than 7,600 sterilizations from 1929 through 1974 based on exaggerated claims that it could eradicate mental illness, genetic defects and social ills.

Once the board was disbanded, legislators shifted the responsibility for ruling on sterilization petitions to the judicial system.

The eugenics board has come under increased scrutiny since a series of stories in the *Winston-Salem Journal* revealed new details. Gov. Mike Easley formally apologized for the program and appointed a study committee that started meeting last week to consider potential reparations to victims, among other things.

It is not clear how many sterilizations have been approved by District Court judges since the 1970s, but Dick Ellis, a spokesman for the N.C. Administrative Office of the Courts, said that such petitions are uncommon.

"Our statistics only show there are three of them done in North Carolina (in 2001-02)," Ellis said. "It's very rarely used."

A spokesman for the N.C. Department of Mental Health and Human Services said that in the last five years, one patient at a state mental hospital had been sterilized.

Womble sees little difference between the sterilizations ordered by the eugenics board and those approved by judges.

"It's a leftover," he said of the existing law. "I want to do all I can to remove all vestiges ... to that kind of program that was done in this state."

Currently, parents or guardians of the mentally challenged - or a head of a state mental institution - can request a sterilization order. The law says that sterilization is designed to protect children from being born to parents incapable of caring for them or to prevent the birth of children likely to suffer "physical, mental or nervous diseases or deficiencies."

"The people of North Carolina have a right to prevent the procreation of children who will become a burden on the state," the N.C. Supreme Court said in 1976 when it upheld the law.

Before a judge can order the procedure, it must be shown that the person is likely to have sex without contraception and therefore likely to get pregnant.

Rep. Verla Insko, D-Orange, who supports eliminating the sterilization law, said that the state should not be involved in decisions of personal rights and freedoms.

"It's really scary that the state is in the business of sterilizing people," she said. "It's not an appropriate role for government."

California is latest state to apologize for eugenics

It 'must never be repeated,' governor says
March 13, 2003

By Kevin Begos
JOURNAL WASHINGTON BUREAU

California has become the fifth state to apologize for its eugenic sterilization program, but North Carolina is still the only state that is examining the possibility of reparations for victims.

Tuesday's statement by Gov. Gray Davis of California was very similar to ones made in December by Gov. Mike Easley and Gov. John Kitzhaber of Oregon. Gov. Jim Hodges of South Carolina issued an apology in January, just before leaving office. Virginia made the first apology, early last year.

"Our hearts are heavy for the pain caused by eugenics. It was a sad and regrettable chapter in the state's history, and it is one that must never be repeated again," Davis said.

Easley appointed a commission in February to investigate how the program unfolded in North Carolina, how to prevent a similar scenario in the future and how to redress the victims.

The commission will meet again Friday. None of the other states have such panels.

After years of silence, the nationwide trend of apologies is being driven by a network of scholars and activists who follow the issue across state lines.

Advocates of disability rights brought up the apologies made by other states for sterilization programs during a meeting with Davis this week, said his spokesman, Russ Lopez.

Davis agreed to the idea, a decision helped by the fact that Paul Lombardo, a eugenics scholar from the University of Virginia, was testifying before a California Senate committee on Tuesday, Lopez said.

California passed a sterilization law in 1909, and it led the nation with more than 20,000 operations. Virginia was second, with about 8,000, and North Carolina was third, with about 7,600.

Most of the California sterilizations were done before 1945, and most in North Carolina during the 1950s and 1960s.

Lopez said that in California, as in other states, most people simply didn't know that such things had taken place in the past.

"When a lot of folks heard of this they were shocked," he said.

James Dunn, a professor of Christianity and public policy at the Wake Forest University Divinity School, welcomed the trend.

"The apologies are very important. They're not the last word, but they may be the first word" in a process, he said.

"They recognize that there is an ethic and morality that is greater than, higher than, anything that the law may allow. The law may have permitted or allowed or even encouraged something that is now wrong," Dunn said. "But the law can't determine whether anything is moral or ethical. It may nudge us in the right direction, but it doesn't have the final say."

Lombardo said he was pleased that other states are apologizing for sterilizations, but that isn't enough by itself.

"The apologies are only the first step. I hope the lack of originality in the language they choose is not an indication that the process is becoming trivialized," said Lombardo.

Kitzhaber said in December, "Our hearts are heavy for the pain you endured."

Davis said, "Our hearts are heavy for the pain caused by eugenics."

Easley said in December, "This is a sad and regrettable chapter in the state's history, and it must be one that is never repeated again."

Lombardo, the director of the program of law and medicine at the University of Virginia's Center for Biomedical Ethics, said, "We will have to wait to see how much follow-up and attention to these issues remains after the initial sentiments are voiced."

The eugenics movement was founded in the 1880s by Charles Darwin's cousin, Francis Galton. It made exaggerated claims that mental illness, genetic defects, and social ills could be eliminated by sterilization. The laws concerning sterilization varied between states, but the U.S. Supreme Court legitimized the basic idea of forced sterilization in a 1927 ruling.

Even critics of eugenics agree that any formal step beyond an apology is complicated by legal, medical, and historical uncertainty. A small number of operations in North Carolina were done because the patient wanted the surgery, and many other records are incomplete, or still under seal in the state archives. Many records in other states have been lost or destroyed.

Dunn said he didn't have a formula for how to proceed on possible reparations. Lombardo said last month he thinks that the only realistic possibility would be to limit payments to the actual victims of sterilization who are still alive - not relatives of people who are dead.

Nationwide about 65,000 people were sterilized, but there are no reliable estimates of how many victims are still alive.

Eugenics panel hears of pain

'It's like a cancer that eats you and eats you and eats you,' victim says

March 15, 2003

By John Railey
JOURNAL REPORTER

RALEIGH – Elaine Riddick Jessie and Nial Cox Ramirez, sterilized more than 30 years ago by order of state officials they never met, finally got to tell their stories yesterday of the pain that those operations caused.

"Yes, I have a lot of anger in me ... because of what I had to go through all my life," said Jessie, who was sterilized in Chowan County in 1968, when she was 14. She told a committee appointed by Gov. Mike Easley how the operation happened after she became pregnant by a man in his 20s - statutory rape by law. "I was only a child," she said.

Jessie and Ramirez cried as they talked, as did some members of the Eugenics Study Committee, which is considering reparations and other forms of compensation for those sterilized by the Eugenics Board of North Carolina. It ordered operations on more than 7,600 people from 1929 through 1974.

"The sadness that I feel about what has happened then turns to rage about has happened," said Carmen Hooker Odom, the chairwoman of the committee and the secretary of the N.C. Department of Health and Human Services.

Easley apologized for the eugenics program in response to a series of stories in the *Winston-Salem Journal*, two of which featured Jessie and Ramirez. The Eugenics Study Committee is the first of its kind in the nation.

Yesterday, committee members asked Jessie and Ramirez what they want the state to do for them.

"I want whatever I can get out of the state of North Carolina," Jessie said. "They should have just cut off my arm or just killed me."

Ramirez said she just wants to make sure that no one else is ever sterilized.

"I would want that no one would ever have to go through the pain that I went through," she said.

She and Jessie talked to attentive state officials yesterday in a conference room not far from the one where the five-member Eugenics Board voted to have them sterilized in the 1960s. They never got a chance to appear before that board.

"Who were these people ... who can go behind closed doors and make a decision on your life, how your life is going to be run?" Ramirez asked.

She and Jessie, who met face-to-face for the first time yesterday, have similar stories. Both come from northeastern North Carolina. They are black and were poor, as were most of those sterilized by the state in the 1960s.

Social workers prepared petitions to have both sterilized, and both were sterilized after having one child. Ramirez said that her social worker threatened to take her family off welfare if she didn't consent to be sterilized. "If I could just get my hands around her neck, I would kill her," she said.

Ramirez was sterilized in 1965 in Washington County when she was 18.

She and Jessie moved to New York, and now live in the Atlanta area. They have little use for North Carolina, and can't forget what the state did to them.

"I tried so hard to bury this, but it just won't go away," Ramirez said. "It's like a cancer that eats you and eats you and eats you."

As one justification for being sterilized, Riddick and Jessie were labeled "feeble-minded" on the basis of flawed intelligence testing. The two children born to the women work in the computer field.

Jessie's son, Tony Riddick of Winfall, and Ramirez's daughter, Deborah Chesson of Georgia, accompanied their parents to the committee hearing. When his mother couldn't talk through her tears, Riddick took over.

"Who has the audacity, or really the authority, to do these things, and where do you get it from?" he asked.

He was glad for the committee's work, he said, then placed his hands on either side of his mother's head. "I just wish we could fix this right here."

Chesson wondered what the children of those sterilized might have accomplished.

"One of those babies could have grown up and found a cure for cancer or diabetes or MS," she said.

Hooker Odom said that the committee hopes to submit in June its recommendations to Easley about what should be done.

Options include help with health-care costs incurred by those sterilized. Jessie said that the operation left her with bleeding spells for years afterward, and said she sees a psychiatrist and takes Prozac as well as another drug to help her sleep.

Jessie and Ramirez came to Raleigh at the invitation of Rep. Larry Womble, D-Forsyth, who is a member of the study committee. On Tuesday, the House Health Committee is scheduled to discuss a bill sponsored by Womble that would erase one of the final remnants of the state sterilization program: a law that gives judges the right to order sterilizations for the mentally ill. The law has rarely been used.

An amendment to the bill would give parents the right to have their mentally retarded or mentally ill children sterilized in special circumstances.

Womble and the committee are also considering other laws that might help those sterilized and make sure that such a program never happens again.

As they explore ways of helping those sterilized, committee members said they plan to ask for comments from Jessie and Ramirez and any other victims who might come forward.

Pam Young and other committee members thanked Jessie and Ramirez for telling their stories.

"You're two very strong and courageous women ... to have endured these kinds of feelings and thoughts for so many years," said Young, the deputy secretary for the N.C. Department of Cultural Resources.

Panel accepts change to law
Womble's bill would end last tie to eugenics
March 19, 2003

By Kevin Begos
JOURNAL REPORTER

RALEIGH–An N.C. House committee unanimously approved a bill yesterday that would repeal a state law that allows for the involuntary sterilization of the mentally ill.

The law is the last legal link to North Carolina's eugenic sterilization program, which authorized more than 7,600 sterilizations from 1929 through 1974.

Rep. Larry Womble, D-Forsyth, sponsored the bill after learning of the abuses that took place under the sterilization program, which was based on exaggerated claims that it could eliminate mental illness, genetic defects and social ills.

Many of the sterilizations were involuntary or based on questionable IQ tests. Some were performed on children as young as 10.

Changing the law will help make sure that such abuses never happen in the future, said Womble, who was joined yesterday by two victims of the sterilization program at a meeting of the House Health Committee.

"Don't let this continue," said one of the victims, Nial Cox Ramirez, who was sterilized in 1965 at age 18 after the birth of her daughter.

"Everybody has a right to have children," said the second victim,

Elaine Riddick Jessie, who was sterilized in 1968 at 14 after the birth of her son.

Ramirez and Jessie left North Carolina shortly after their operations, and both now live in the Atlanta area.

The law is seldom used. Three petitions were granted in 2001-02. Over the past five years, only one patient at a state mental hospital was sterilized.

The state's eugenic sterilization program has come under increased scrutiny since a series of stories in the *Winston-Salem Journal* in December exposed previously unreleased details.

Gov. Mike Easley formally apologized for the program and appointed a committee to investigate how it happened, how to prevent a similar scenario in the future and how to compensate the victims.

The law that Womble's bill would repeal allows sterilization in cases of a person's "mental, moral or physical improvement," or for the "public good."

"The people of North Carolina have a right to prevent the procreation of children who will become a burden on the state," the N.C. Supreme Court said in 1976 when it upheld the law.

The final version of Womble's bill includes an amendment giving parents the right to have their mentally retarded or mentally ill children sterilized in special circumstances.

The guardian of a mentally ill or mentally retarded person would be able to petition a clerk of the court for an operation by providing a sworn statement from a doctor that it is medically necessary and is not for the sole purpose of sterilization.

A sworn statement from a psychiatrist or psychologist to determine if the person is able to comprehend the nature of the operation also would be required.

Womble's bill will now go before the full House for a vote and then to the Senate, if it passes.

Easley supports the bill, Womble said.

House votes 116-1 to end sterilization law
Rarely used law a remnant of N.C. eugenics program

By Dana Damico
JOURNAL RALEIGH BUREAU
March 27, 2003

RALEIGH– With no debate, the N.C. House voted 116-1 yesterday to strike a law that allows involuntary sterilizations of the mentally ill.

Though rarely used, the law remains the last legal link to the state's eugenic sterilization program that ordered more than 7,600 sterilizations from 1929 to 1974, many of them against the wishes of the patients and their families. Rep. Larry Womble, D-Forsyth, pushed to have the law repealed after he learned of the abuses of the eugenics program, which was based on flawed claims that sterilization could eradicate mental illness, genetic defects and such social ills as out-of-wedlock births.

Recent reports also revealed that many of the sterilizations performed were based on questionable IQ tests and without the patients' consent.

Some were performed on children as young as 10.

"It's unconscionable," Womble said. "It's ungodly. It's not the American way."

The proposal now moves to the Senate for consideration.

Gov. Mike Easley has formally apologized for the state's eugenics program, and appointed a committee to investigate and consider ways to compensate victims. Two weeks ago, the panel heard emotional testimony from two women who were sterilized in the mid-1960s.

Both women are black and were poor. Social workers prepared petitions to have them sterilized after each had a child.

Elaine Riddick Jessie was 14 when she was sterilized. She got pregnant by a man in his 20s - statutory rape by law - and was labeled "feeble-minded." Nial Cox Ramirez, who was 18 when she was sterilized, said that her social worker threatened to cut her family from welfare if she didn't agree to the operation. She, too, was deemed "feeble-minded."

Both gave birth to children who work in the computer field.

Womble said that the testimony was heart-rending. And he said that there are other victims like Jessie and Ramirez who have not stepped forward but have suffered silently the shame and heartbreak of being forcibly sterilized.

"We should compensate these people," he said after the debate. "My next priority is to work on reparations."

The bill passed yesterday would repeal the law that allows District Court judges to approve involuntary sterilizations for a person's "mental, moral or physical improvement," or for the "public good."

The law took effect in 1975 after legislators dissolved the eugenics board. Once the board was disbanded, legislators shifted the responsibility for ruling on sterilization petitions to the judicial system.

"The people of North Carolina have a right to prevent the procreation of children who will become a burden on the state," the N.C. Supreme Court said when it upheld the law in 1976.

Three sterilizations were granted in 2001-02. State health officials say that in the past five years only one patient at a state mental hospital has been sterilized.

The bill would still allow involuntary sterilizations in certain instances, including if a mentally disabled woman needs a hysterectomy to remove fibroid tumors or treat ovarian cancer.

Under the legislation, the guardian of a mentally ill or retarded person could petition a clerk of court for the operation. They would have to provide a doctor's sworn statement that the operation is medically necessary and not solely for sterilization.

After the vote, the lone dissenter, Rep. Russell Capps, R-Wake, said he might not have understood the bill. "Once I pushed the button, I tried to turn it off and I couldn't," Capps said. "I wasn't voting for sterilization because I don't know if there's a need for it."

Capps did say, however, that there could be instances when someone should be sterilized, including when severely mentally retarded people have children they can't care for.

"I might of made a goof," he said. "I don't know. It won't be the first one."

Senate votes to repeal sterilization law
Easley expected to sign bill; panel to look at reparations
April 4, 2003

By David Rice and John Railey

RALEIGH– The N.C. Senate unanimously agreed yesterday to remove one of the last remnants of a state eugenics program that sterilized more than 7,600 people from 1929 through 1974 - often against their will.

The Senate voted on a bill that the House overwhelmingly approved last week, one that would remove a law allowing the involuntary sterilization of the mentally ill.

"Obviously, I was thrilled and feel very grateful that both chambers understood the enormity of this situation and acted quickly and responsibly," said Carmen Hooker Odom. She is the chairwoman of a committee appointed by Gov. Mike Easley to consider reparations and other forms of help for those ordered sterilized by the Eugenics Board of North Carolina.

Hooker Odom, the secretary of the N.C. Department of Health and Human Services, said that the Senate action clears the way for the committee's final work: helping those sterilized by the program. The eugenics board has come under increased scrutiny since a series of stories in December in the *Winston-Salem Journal*.

Easley formally apologized for the program on the last day of the series then appointed the study committee, the first of its kind in the nation. Easley is expected to sign the bill approved into law soon, Hooker Odom said.

The law to be stricken allows sterilization in cases of a person's "mental, moral or physical improvement," or for the "public good." It took effect in 1975 after legislators dissolved the eugenics board, which was driven by aggressive social workers and often relied on flawed IQ tests. The program - based on the eugenics movement's claims that sterilization could eradicate mental illness, genetic defects and social ills - operated with little oversight.

"Bureaucracies sometimes get these lives of their own and just keep on going, even if people don't realize they're still in existence," Hooker Odom said.

Sen. Jeanne Lucas, D-Durham, who spoke for the bill, pointed to emotional testimony before legislative and the governor's committees recently from two women who had been sterilized against their will in the 1960s in northeastern North Carolina. The women, Nial Cox Ramirez and Elaine Riddick Jessie, both live in the Atlanta area now.

"These women have been able to carry on their lives in spite of what happened to them. And their children - their first set of children - have led productive lives," Lucas said.

Rep. Larry Womble D-Forsyth, the bill's principal sponsor, said he hopes that Easley will have a ceremony to sign it and invite the women to attend.

Once the eugenics board was disbanded, legislators shifted the responsibility for ruling on sterilization petitions to the judicial system. The new bill still allows for involuntary sterilizations in certain narrow circumstances, including if a mentally disabled woman needs a hysterectomy to remove fibroid tumors or treat ovarian cancer.

After the vote, Sen. Ellie Kinnaird, D-Orange, said that North Carolina's long experiment with eugenics showed that science can indeed be mistaken.

"Society sometimes can be very, very wrong. And we look sometimes on the despised of society as something we can do whatever we want with," Kinnaird said. "The lesson I hope we learn from this is that we treat everyone with respect."

Backers of the bill say that the next step is to push for reparations for victims.

The committee appointed by Easley is studying the issue. But given the state's current budget troubles, some aren't sure that the state can afford to pay victims.

"I've got that bill drafted, ready to go," Womble said. "It doesn't have to be money, either. It could be educational considerations. It could be health - some of these people are still suffering health effects." Lucas said that Easley's committee is studying the possibility of reparations. "Whether it can be done or not, I don't know. But it certainly needs to be looked into," she said.

Others fear legislative inaction, saying that reparation payments could raise another, more controversial issue.

"Lurking behind those reparations are the biggest reparations of all, for slavery," said Kinnaird. But she said that the United States has paid reparations to Japanese-Americans who were held in internment camps during World War II.

Hooker Odom said that considering compensation for those sterilized isn't "in the same discussion as full reparations around issues related to slavery." The difference, she said, is that the state eugenics program involved a limited number of victims sterilized by a single program, she said.

Her committee hasn't taken financial compensation for those sterilized out of consideration, she said, but it's also considering other options, such as health-care credits and education credits. The committee hopes to submit in June its recommendations to Easley about what should be done.

Easley wanted it to move in a timely fashion in helping those sterilized by order of the eugenics board, Hooker Odom said.

"I think he is appalled that it happened here in North Carolina," she said. "And as long as it happened in North Carolina, and the direction it took in North Carolina."

Class played a role in eugenics sterilizations, researcher says
April 11, 2003

By Theo Helm
JOURNAL REPORTER

People were targeted by the state's eugenics program because of their class and gender as much as their race, a professor of women's studies told a group at Wake Forest University yesterday.

"Racism was part of the problem but not the whole problem," said Johanna Schoen, whose doctoral work on the Eugenics Board of North Carolina played a key role in "Against Their Will," the *Winston-Salem Journal's* series about sterilization.

The eugenics board ordered the sterilizations of more than 7,600 people from 1929 through 1974, often against the wishes of the women and their families.

The board sterilized increasing percentages of black women as it progressed, but Schoen said that it is important to not forget about how the program took advantage of poor women.

The eugenics program showed that a great number of people appeared to believe that women shouldn't have children while they are on welfare - attitudes that were reflected in some of the reactions to the Journal's series, Schoen said. It goes along with some people's belief that some women have more children to receive more welfare benefits, she said.

"I would argue that public attitude has changed very little over the decades," Schoen said. As examples, she pointed to President Clinton's welfare reform that instituted financial penalties for women on welfare who have children and President Bush's emphasis on abstinence education. "Rights are only as strong as we are willing to tolerate the choices of others, even if we disagree with them," Schoen said. Gov. Mike Easley apologized for the eugenics program in December and appointed a committee to consider reparations and other forms of compensation for those sterilized. The committee hopes to submit its recommendations to Easley in June.

Easley is expected this week to sign a bill that would remove a law allowing the involuntary sterilization of the mentally ill.

Another committee - this one appointed by Dr. William Applegate, the dean of the Wake Forest University School of Medicine - is looking into the university's role in the eugenics program. "We're generally focusing on what if any institutional role was there in this," Applegate said.

A "very experienced faculty member" is leading the committee, Applegate said. Applegate declined to say how many people are on the committee or the members' names.

The committee should be done sometime this spring, and will send the report to him, Applegate said.

Easley repeals eugenics statute

Two women sterilized under old law attend signing ceremony
April 18, 2003

By Dana Damico
JOURNAL RALEIGH BUREAU

RALEIGH – More than 30 years after Elaine Riddick Jessie and Nial Cox Ramirez were sterilized by state officials they never met, the law that authorized the irreversible operations was officially repealed.

Jessie and Ramirez, who traveled from Atlanta to Raleigh to celebrate the occasion, were introduced as "brave and courageous ladies who stepped forward in order to get that bill passed." Joined by their families, the two women received a standing ovation from members of the N.C. House of Representatives and onlookers in a packed balcony at the Legislative Building.

"I want you to let them know that North Carolina is not that North Carolina that forced sterilized them years ago," Rep. Larry Womble, D-Forsyth, told his colleagues as he introduced them. "We are better than that today."

Womble led the legislative effort to repeal the involuntary sterilization law - the last legal link to the state's eugenics program that sterilized more than 7,600 people from 1929 to 1974, many of them against their will.

Gov. Mike Easley signed the bill into law at a private ceremony with Jessie and Ramirez, their families and several legislators, including Womble.

"It was a very joyful event to see that someone actually took the time and heard our cries ... took the time and paid attention," Jessie said. Her only son, Tony Riddick of Winfall, and her nephew, Curtis Riddick of Georgia, joined her.

"No one should ever feel the pain and agony of not being able to have children," Jessie said. "It's a God-given right: Be fruitful and multiply and fill the world with images of thyself. This is something they took away from us."

Jessie and Ramirez were sterilized by order of the Eugenics Board of North Carolina. The board operated with little oversight, justified sterilizations on flawed intelligence testing and often ignored the wishes of patients and their families.

Its work went largely unscrutinized until Johanna Schoen, a graduate student at the University of North Carolina at Chapel Hill, got access to previously sealed records from the eugenics board.

Easley apologized for the program after a series published in the *Winston-Salem Journal* revealed new details. He also created a special committee, the first of its kind in the country, to study whether victims should be compensated. The committee will meet again April 24 and could make recommendations by June.

The law that was struck allowed involuntary sterilizations "for the public good" or when officials thought it likely that the person would have a child with a physical or mental disease. It was a remnant of the law that authorized the eugenics board to order the operations on people it deemed mentally ill, diseased or feeble-minded.

Jessie was 14 when she was sterilized. The board determined that she was "feeble-minded" after she got pregnant by a man in his 20s, statutory rape by law. Ramirez was 18 and the mother of one when she was sterilized. She, too, was labeled "feeble-minded." She said her social worker threatened to cut her family from welfare if she didn't agree to the operation.

Both women have agonized about the sterilizations for years. They say they're happy that no one else will have to suffer the same feelings now that the law has been repealed - only medically necessary sterilizations can be performed now. But they say that the official bill signing signals a new beginning, not the end of a bad chapter.

"I wanted him (Easley) to say what else he was going to do for us. He didn't," said Ramirez, who was accompanied by her daughter, Deborah Chesson of Georgia.

Like Jessie, Ramirez says that victims of the eugenics program should be compensated with more than the pens Easley gave them after the signing.

"Signing of this bill is fine," Jessie said. "Thank you very much. But we lost something. It's like losing a limb. You can't use it anymore.... It ain't about gold-digging, it's the principle."

Tony Riddick said that his mother would never be the same. She sees a psychiatrist and takes Prozac as well as another drug to help her sleep.

"My mother's sick. I know why she's sick," he said. "Whenever I see her sick, I think of the wrong that was done to her.

"I think they should continue this process," Riddick said. "It shouldn't stop here."

Chesson, too, expressed hope for her mother's future, despite the renewed pain that comes with such events.

"I think it can be a beginning to healing," she said. "But if it was all just a show, then they're actually being victimized again."

Redress, counsel is aim of project
Bill would create panel to study compensation for sterilization victims
April 24, 2003

By Dana Damico
JOURNAL RALEIGH BUREAU

RALEIGH– Under legislation filed yesterday, North Carolina could become the first state to compensate people involuntarily sterilized as part of a nationwide eugenics movement.

The bill - filed by Rep. Larry Womble, D-Forsyth - would create a legislative research commission to determine how to compensate and counsel victims.

It assumes that the General Assembly agrees that those among the more than 7,600 people sterilized from 1933 to 1974 who were sterilized against their will should be compensated.

And it stops just short of requiring the General Assembly to create a compensation program, regardless of the commission's findings.

"It is the goal of the General Assembly after reviewing the results of the commission's study to create and fund a program during the 2004-05 fiscal year to compensate, in some form, and counsel persons who were sterilized through the state's eugenic sterilization program," the bill reads.

"I know a great many people are talking about money," Womble said. "But it's not limited to just the finances. It's also education and health and any other kinds of concerns. That's what we're working toward."

The bill aims to redress the wrongs of the state's eugenics program - the third largest in the country, after California and Virginia.

Nationwide, about 65,000 sterilizations were performed as part of the eugenics movement. It sought to end mental illness, genetic defects and social ills thought to be passed down from parent to child.

Gov. Mike Easley apologized in December for the state's involvement in the program after a series in the *Winston-Salem Journal* gave details about the workings of the Eugenics Board of North Carolina.

Easley then appointed a panel to examine the program and consider compensation. The group, which includes Womble, has met several times since February and heard emotional testimony from two women who were sterilized against their will. The panel is scheduled to meet again today.

Womble wants to create a separate commission because he said that it's not certain what the panel will suggest.

"It's not guaranteed that there will be some kind of reparation or compensation," Womble said.

Carmen Hooker Odom, the state secretary of health and human services who heads the panel, supports efforts to compensate victims and has suggested that the state could offer health coverage or counseling. But she warned that it would be hard to find victims and families.

State officials cannot review all the documents related to the eugenics program because of confidential medical information in them.

Panel calls for compensating N.C. eugenics victims
Easley's commission issues recommendations
to identify, help those who were sterilized
May 31, 2003

By Dana Damico
JOURNAL RALEIGH BUREAU

RALEIGH–After becoming the first state to create a commission to review its role in a statewide eugenic sterilization program, North Carolina could soon take another dramatic step to compensate unwilling victims of the program.

A national media campaign to identify and compensate victims could start in August under recommendations tentatively endorsed yesterday by the Eugenics Study Committee.

Gov. Mike Easley appointed the group to scrutinize the Eugenics Board of North Carolina, which authorized more than 7,600 sterilizations from 1929 until 1974, and to consider how to redress those people who were forcibly sterilized.

The study committee plans to submit its report to Easley in coming weeks.

The report calls for the creation of a special health fund to provide medical care for victims, and free education benefits at the state's universities and community colleges for victims and, potentially, their families.

It also backs the concept of reparations supported by Rep. Larry Womble, D-Forsyth, a leading critic of the state's role in the eugenics movement.

"Although the committee strongly believes that survivors also deserve some form of financial compensation for what we believe is a violation of human rights, we recommend the creation of a legislative study commission to address this issue," the report reads.

A bill to create such a commission is pending in the General Assembly.

The committee has met four times since February. In March, it heard heart-rending testimony from two women who were sterilized against their will.

One was Nial Cox Ramirez, who was sterilized in 1965 when she was 18. Ramirez, who has one child, said that her social worker threatened to drop her family from welfare if she didn't undergo the operation.

Ramirez's story is featured in the current edition of Newsweek magazine.

Nationwide, about 65,000 people were sterilized as part of the eugenics movement. Supporters of eugenics wanted to eliminate such social ills as out-of-wedlock births, genetic defects and mental illness thought to be hereditary.

North Carolina's response to the eugenics program has drawn national and international attention. Easley formally apologized for the program in December after a series of stories was published in the *Winston-Salem Journal*. Since then, the General Assembly voted to strike a law that allowed for the involuntary sterilization of the mentally ill - a final legal vestige of the eugenics program.

The committee report calls North Carolina's sterilization program a "shameful blot" on state history that should never be repeated.

If Easley accepts the recommendations, efforts to find people sterilized against their wishes could start in July with a statewide news release. Privacy laws bar state officials from identifying victims from the more than 150,000 pages of documents stored at the state archives on the program. Those records are locked up, officials said.

Budget constraints also prevent a costly publicity campaign and committee members suggest the use of "free media."

To find victims, it recommends posting information in church bulletins, N.C. Division of Motor Vehicles offices, health departments, libraries and schools, as well as on billboards or even city buses. It recommends enlisting help from hospitals and doctors, faith-based groups and churches, civic organizations and the National Association for the Advancement of Colored People.

"This is going to be a national thing," Womble said. "Hopefully, those who want to, will find it comfortable enough to come forward." Committee members said that they have no idea how many people that could be. Many people may have died or moved across the country. Some may not want to revisit painful memories of the irreversible operation. Others may be skeptical of getting redress from a state that violated them.

"My hope is that we'll have a groundswell," said Carmen Hooker Odom, the secretary of the N.C. Department of Health and Human Services and the head of the study committee. "We're hoping to get folks to come forward so we can do the right thing, finally."

According to news reports, state officials already have received requests from people who say they were forcibly sterilized. Womble says he has a list of 12 people. Another 12 have called a state hot line.

Cathy Morris, the state archivist, said that 10 people have contacted her office, including Ramirez and another victim, Elaine Riddick Jessie.

Jessie, who joined Ramirez in testifying before the committee, was 14 when she was sterilized in 1968 after a man in his 20s impregnated her - statutory rape by law.

Of the other eight, only one had records on file at the archives. "Which is an interesting commentary," Morris said.

Under the committee's draft plan, people who believe that they were sterilized would request their records from the state archives. They would be eligible for benefits if a three- or five-person panel certified their claim.

The committee agreed to start work immediately on a survivors' group. Members could look for emotional support from one another and aid other victims through the potentially intimidating process, the committee said.

"They're very upset," said Debbie Crane, a health and human services spokeswoman who has talked to victims. "It is tough trying to talk about it."

Womble said that it was a monumental decision for Ramirez and Jessie to share their stories. Officials need to recognize that others may be equally fearful. "We must be almost like a confidante to them," he said. "Whatever it takes to hold their hands."

"They've been carrying this burden."

Offer 'Too Little Too Late'

Woman says threats forced her to agree to operation and that she expects N.C. to provide compensation
September 28, 2003

By John Railey
JOURNAL REPORTER

Alone with haunting memories in her Connecticut home, Ernestine Moore spent much of last winter waiting to hear what Gov. Mike Easley would do for her and thousands of others sterilized by the state of North Carolina from 1929 through 1974.

This week, she heard what the state was offering -mainly, education and health credits. For Moore, it was too little too late.

"That's all they're offering?" she asked from her Bridgeport home. "I'm not satisfied with that. They really messed up my life," said Moore, who was sterilized in Pitt County in 1965 after having her first and only child. She was 14.

Moore wants money for her pain.

Ted Gartman Jr. of Greenville, who signed off on Moore's sterilization as the local director of public welfare, isn't sure that financial compensation is the answer. He said that the sterilization program wasn't one of "unfettered power," of someone just sitting down and deciding to have a male or female sterilized.

"It may have been misused.... I'd like to think in our agency it wasn't, but it's conceivable it could have been," said Gartman.

Her long silence

In 1965, Moore was living with her mother and seven brothers and sisters in a four-room house in Fountain, a small town outside Greenville in Pitt County.

She got pregnant by a boyfriend, she said. She remembers a visit to a clinic, where a white social worker encouraged her to be sterilized after the birth of her child. "I really didn't want them to do that," she said. "They told me if I didn't do it, my mother would be cut off welfare. If I didn't do it, they'd take my child away and put it in a home. I had to do something."

Both Moore and her mother signed a consent form for her to be sterilized.

The petition for the operation included test results that classified Moore as feeble-minded.

"If I were feeble-minded like they said, seemed like they would have put me in an institution," she said.

She gave birth to a daughter, Sharon Diane Moore, on Aug. 29, 1965, at Pitt County Memorial Hospital. The next day, she said, she was sterilized.

Moore said she remembers hearing doctors talking during the operation.

"I heard them saying they really didn't want to do it. They said I was too young for them to do it."

No one told her the operation was irreversible, she said, and she received no counseling in the days after it.

Two years after the operation, Moore and her daughter moved to Bridgeport, where friends from North Carolina were already living. She worked as a nurse's aide, she said, and in factories.

She said she entered into a relationship and wanted to have another baby, but couldn't get pregnant. Her doctor sent for her records from Greenville, she said, and that's when she learned her sterilization was irreversible.

Like others sterilized, she didn't talk about it. "I kept that secret from the world," she said. "I was kind of ashamed of it."

Moore said she never told her daughter about what happened. Police say that Sharon Moore committed suicide by hanging herself in 1995. Moore believes that her daughter was murdered.

Ernestine Moore, divorced, is now alone, save for visits from her daughter's two grown children.

"I don't have a daughter no more and I couldn't have no more kids. See, I don't have nobody."

Confronting the past

This winter, Moore realized she wasn't alone in her pain after a friend in North Carolina sent her a newspaper story about a committee set up to study the state sterilization program, and stories about the testimony that others had given.

Moore sent off for records from the state archives in Raleigh and a few weeks later she got a letter from the state. She tore it open, finding copies of typewritten forms about her that she had never seen. The words describing her at 14 chilled and angered her.

On one page, social worker Jo Ann Smith wrote that it was hard to discern Moore's feeling about being sterilized. "(Moore) is so withdrawn that worker has been unable to detect any emotion other than fear."

On another page, Smith wrote that most of the people in Moore's neighborhood were "of low incomes and low morals." Eugenics-board records are filled with similar petitions, ones that read more like gossip than sociological analysis.

"A whole lot of them things they said were not true," Moore said. "She didn't know nothing about them people."

Included in the petition was a test report that said Moore had an IQ of 56. But the report also said that "Ernestine has no appearance of retardation" and "the exact test score is undoubtedly a minimal estimate of her potential."

As in numerous other state sterilization cases, the psychologist who gave the IQ test did not refer to feeble-mindedness. In Moore's case, as in many others, a doctor made a check by "feeble-minded" on a form in the petition. Dr. Harold Hoke wrote on the form that he had just met Moore the day he filled out the form.

As Moore read the words, she said, she decided it was time to break her silence, set the record straight and seek compensation for what happened.

Unfinished business

Gartman, 66, went on to teach social work at East Carolina University in Greenville before retiring. Like many of those involved in the sterilization program, he has mixed feelings about it.

Sterilization in the South was probably "something of a poverty control thing" aimed primarily at blacks, he said, but whites were sterilized as well.

"The community feeling that I got was 'We've got these people on welfare and they keep having children and we've got to do something about it.'"

Yet, he said, some parents, including those of mentally handicapped children, asked that they undergo the operation.

"In those days, we didn't have the pill.... Believe it or not there were parents and relatives who came to our agencies begging for this."

There might be people who deserve compensation for being sterilized, he said, but "I have a hard time saying that anybody who was sterilized ought to be compensated for it.... I don't think money is the answer to everything."

He doesn't remember Moore's case.

"The only thing I can say is I regret if the system was flawed in any way," he said.

Moore says that the sterilization program was systemically flawed and the state should pay for her suffering. She said she plans to get a lawyer to pursue reparations and to speak out about what happened to her.

"I'm not finished with this. And they did that to a lot of people."

Suggestions abound; Wheels turning slowly
September 28, 2003

By Danielle Deaver and John Railey
JOURNAL REPORTERS

How do you compensate those who were sterilized by the state of North Carolina? Women who were sterilized, legislators, members of a study committee established this year by Gov. Mike Easley and the governor himself have spent months trying to answer the question.

In a national first, they have come up with a list of recommendations – approved by Easley last month – that they hope will ease the pain suffered by those sterilized and at the same time provide reminders and education to prevent a similar tragedy. Most of the victims were poor women who were often coerced into sterilization by social workers. Inaccurate labels of "feeble-mindedness" were often used as justification.

Some of the proposals would directly help the victims. The committee recommended providing education benefits through the University of North Carolina system and community colleges. It also recommended setting up a special fund to provide health care. The benefits would be provided possibly through Medicaid or the state health plan.

Committee members want to set up a nonprofit group to help find and support those sterilized. They also want to establish a system in which those sterilized can get help, possibly from college students, in negotiating the maze of medical records needed to confirm their stories.

Committee members also recommended building a memorial to the more than 7,600 people who were sterilized by the state and including information about the program in the state's North Carolina history curriculum. They also want to hold a seminar to discuss the eugenic sterilization program with experts. And the N.C. Department of Health and Human Services will establish a course in ethics that will be required for all professional workers in the department.

"Anytime we take away a person's ability to choose, there has to be a great deal more study and evaluation than we used in the past," said Ted Gartman Jr. of Greenville, who signed petitions for sterilizations as the director of public welfare in Pitt County from 1965 through 1969.

In response to a series of stories published in the *Winston-Salem Journal*, Easley apologized in December for the actions of the Eugenics Board of North Carolina, which ordered the operations from 1929 through 1974.

Easley then did what no one else has done in the other 32 states that had similar programs – he established a eugenics study committee to consider reparations and other forms of compensation for those sterilized and approved its recommendations.

The money for some of the programs, such as the health credits, will have to come from a special appropriation from the General Assembly in 2004. Others, such as changes to the curriculum in public schools, will have to be implemented by other state departments.

Several victims say that reparations are needed as well. They are not happy with the way Easley handled the process, making the decision to back the recommendations quietly in August. Neither the public nor those victims who had already come forward were notified.

Nial Cox Ramirez, one of the women sterilized by the state, recounted her experience to the committee in March.

Later, she visited the governor's mansion to watch Easley sign the bill that repealed the last remnants of the state's sterilization statute. She expected the governor's office to notify her when Easley responded to the committee's recommendations.

"He (Easley) made a decision," said Ramirez, who was sterilized in Washington County in 1965 and now lives outside Atlanta. "He didn't even tell anybody that he made a decision. If you could call us to Raleigh (for the earlier visit)... he could have told us in a formal letter that he did all he could do," she said. "But like I said, forget that. I'm going to let God take care of that.... God is bigger than me."

In its report to the governor, the eugenics committee said it "strongly believes that survivors also deserve some form of financial compensation for what we believe is a violation of human rights." But committee members also said that a legislative study commission would be a more appropriate group to decide about compensation.

State Rep. Larry Womble, D-Forsyth, has introduced a bill that would create such a commission. It would report to the General Assembly during its 2004 session.

Some observers say that reparations are unrealistic in a time when the state is dealing with falling revenues, climbing unemployment and the long recovery from Hurricane Isabel. But Womble said he is "keeping hope alive."

"I'm not so naive as to not realize that there are other priorities, but at the same time this should take top priority ... because this was something that was done against human beings against their will," he said. "I think we do need to make some kind of tangible compensation to these victims."

Making Amends

State struggled to arrive at a consensus about what to do
for those it sterilized and to keep if from happening again
September 28, 2003

By Danielle Deaver
JOURNAL REPORTER

RALEIGH – The scar from Annie Buelin's sterilization surgery still hurts 51 years after the Eugenics Board of North Carolina ordered her to undergo the operation when she was 14.

She is one of more than 7,600 people sterilized by the state from 1929 through 1974 after rudimentary and inaccurate tests indicated that they were "feeble-minded" or had low IQs. Some were sterilized because they had children out of wedlock or reputations for engaging in premarital sex.

Buelin - who still lives near the house where she grew up in Surry County - was not sterilized for any of the reasons that social workers usually gave, though. She was sterilized after her mother gave birth to her 10th and 11th children - both born out of wedlock. Even though Buelin had not had any children and was not sexually active, she was scheduled for surgery with her mother's approval.

The surgery left her devastated and with a string of medical problems that may be related to the procedure. Because of that, her nephew would like to see the state compensate the victims with cash.

"I hope she gets some compensation. I do," Hersie McMillian said.

But Buelin is not sure if the state could ever do enough to make up for what it took.

"That won't change a thing," she said about compensation. "There's nothing that's going to be done, but there should be more than a verbal apology."

Buelin was 14 when a social worker met her at the end of a dirt road near her house. The social worker said that they were going to the hospital.

Buelin asked why.

Surgery, the social worker said. "You have appendicitis."

Buelin had never heard of it, and she didn't seem to have much choice anyway. She had the surgery. Soon afterward, she was home and back to her usual routine of helping her mother carry water, work in the vegetable garden and take care of the family's animals.

Buelin didn't find out what had really happened until several years later. A pain in her side sent her to the hospital when she was 21. Doctors found and removed a cyst from one of her ovaries.

After the surgery, Buelin found out something else that would hurt much more than the cyst.

"They said I had my tubes tied," she said. "My doctor told me, asked me when I had this done. I said I didn't know when I had this done."

Tubes were tied, cut

Her Fallopian tubes had been not just tied, but also cut. Buelin realized that she would never have children.

"It was just like my heart was going to come out of my body. It felt like my heart would split open," she said. "It looks to me like they would have thought, 'This young-un might want to have children someday; why not just tie them?'"

Her husband at the time, Alvin Flippin, told her that he didn't mind that she couldn't have children - he didn't want any anyway. They talked about adopting, but he didn't want that, either.

Buelin let the matter drop. But she thought about what had been taken from her.

"I just think a pregnant woman is the prettiest thing I ever seen. They look miserable but they're pretty," she said. "One of the hardest things that has bothered me about this thing, aside from not having children, was when people would talk about their children or grandchildren."

But Buelin grew up in a family that knew how to do without. Her mother married a much older man when she was only 15 and started having children right away. Two died, one stillborn and one shortly after birth. She had her 11th when she was 33.

The children helped out with everything. They churned butter and tended large vegetable gardens and 16 acres of tobacco. They carried water from the spring.

There was no electricity, so the children cut wood to do everything from cook their food to cure tobacco. Everything was valuable. Hersie McMillian, Buelin's nephew, has a quilt that his grandmother made from the twine that bound leaves of tobacco.

"It was from sunup to sundown," Buelin said. "I thought when mama was successful enough to get a refrigerator that we were rich."

Her older sister, Willie Mae, moved out before Buelin was 14. Her mother allowed Buelin's youngest sister, Betty Jane, to move in with a local minister.

"Betty Jane was just a bookworm and mama just didn't have the means to send her to school," Buelin said. "My mama let a preacher and his wife take my baby sister and raise her and send her to school.... It was so sad."

That left just Buelin and her next-younger sister at home when the social worker came.

Buelin thinks that her mother's actions might have drawn attention from the county department of public welfare.

After Elsie Woods gave birth to her ninth child, her husband developed dementia and went to live in a home in Macon, Ga. He died two years later.

After he left, Elsie Woods became pregnant with another man's child, a boy. After his birth, a social worker approached Woods, Buelin said

Sterilization may have been a condition of receiving welfare; Buelin isn't sure. She only knows that one morning, her mother told her to go to the end of that dirt road and wait.

Her mother gave birth to another illegitimate child, another son, the next year, Buelin said. Then, Buelin thinks, the eugenics board had Elsie Woods sterilized.

"They did mama's in '52 when Charles was born. I knew after he was born she had to go back to the hospital," Buelin said. "She was probably glad."

Buelin was the only child in the family to be sterilized. She thinks that her younger sister's stubbornness saved her from the same fate.

A difficult conversation

Later in her life, Buelin gathered the courage to talk to her mother about the ordeal.

"I asked mama if she gave them permission to do this and she said absolutely not," Buelin said.

But she had. A paper that her nephew found among his grandfather's papers showed that Woods had given permission for the state to sterilize her daughter.

Buelin said she doesn't hold it against her mother.

"She did the best she could. She had all nine of us," she said. "I wouldn't put my mama down. She's like me; she didn't have a lot of education."

Buelin spent her life working in textile mills. The industry's history in this state can be charted through a list of the places she worked - Brown and Wooten, Renfrow, Kentucky Derby, Spencer's.

She also worked at Wayne's Poultry Farm for a while. That's where she earned her nickname, Ducky, because of the way she looked wearing large boots over her shoes.

Her first husband died of cancer in 1972. She remarried 12 years later, to Woodrow Buelin. They are still married.

With Woodrow, Buelin acquired seven stepchildren.

"There's nobody in the world who loves their stepmother more," Hersie McMillan said. "If you ever saw her with a bunch of children. It makes you ashamed - makes you ashamed of your government."

She is retired now and does not have health insurance. She tried to find some that would be affordable, but her health history - she's had two heart attacks - made that impossible.

She has had other problems that she thinks could be related to her unwanted tubal ligation. Seven years after she went through menopause at age 45, Buelin began menstruating again. She had to have a minor surgical procedure to take care of it. There was the cyst. And the scar that she still puts Vaseline on, every day.

And there are other things, Buelin said.

"I do think it's had something to do with my health because I'm real easy to get upset, I get real nervous," she said.

Her stepchildren and many of her friends still don't know what happened to Buelin, or Annie, as her family calls her. Buelin said she's ashamed of having been sterilized.

"I'm sure they probably all wonder why I never had children," she said.

WFU medical school apologizes again for role
Officials criticized in choice of a supporter
November 4, 2003

By Danielle Deaver
JOURNAL REPORTER

After 10 months of study, Wake Forest University School of Medicine issued a report yesterday about its role in North Carolina's eugenic sterilization program, and repeated its apologies for its involvement.

"None of us like what happened here and doubly regret that our institution was involved," said Dr. William Applegate, the dean of the medical school.

The report was critical of a decision by medical-school officials to accept money from a controversial figure who supported eugenics and segregation. "It was bad judgment," Applegate said.

The four members of the committee studied material from the school's archives, records from the Human Betterment League - a group that pushed for the expansion of the state's eugenic sterilization program – minutes from the school's Board of Trustees, hospital records and other material to produce the seven-page report.

The committee, composed of two doctors, a lawyer, and a researcher, started looking into Wake Forest's role in response to "Against Their Will," a series of stories published in the *Winston-Salem Journal* in December. Applegate directed the school to examine its role as soon as he learned that it was involved.

The Eugenics Board of North Carolina ordered the sterilizations of more than 7,600 people from 1929 through 1974. Many of the operations were done against the patients' wishes, and some were performed on children as young as 10. Documents obtained by the Journal showed that North Carolina expanded its eugenic sterilization program while most of the other states with similar programs were scaling back and that the North Carolina program increasingly targeted poor, black women as it grew. Gov. Mike Easley formally apologized for the program in December and approved the recommendations of a study panel in September that called for health and educational benefits for sterilization victims.

Nationwide, about 65,000 people were sterilized as part of the eugenics movement. Supporters of eugenics wanted to eliminate such social ills as out-of-wedlock births, genetic defects and mental illness thought to be hereditary.

The series also examined Wake Forest's involvement with the state sterilization program.

Though the school was involved on a relatively small scale, the Journal investigation showed that:

• Wake Forest accepted money from Wickliffe Draper, a philanthropist with known racist views;

• participated in a eugenic sterilization program along with local elected officials that may have operated outside the purview of the state eugenics board;

• and that the expansion of the state eugenic sterilization program was supported by Dr. C. Nash Herndon, then the head of the Department of Medical Genetics at Wake, in his role as the president of the Human Betterment League.

Draper gave two $40,000 grants to the school in 1950 and 1951 after Herndon presented ideas for genetic research to him. Herndon and Dr. Coy Carpenter, then the dean of the medical school, went to Draper again in 1951, asking for more money for an institute for the study of genetics. Draper agreed to give $100,000 if the school agreed not to officially advocate interracial marriage, to consider teaching about therapeutic sterilization and not to dispute the theory of overpopulation leading to food shortages unless scientific data disproved the theory.

Carpenter agreed to the conditions and accepted the money in 1953. The money was used to pay for Herndon's position, and after his retirement the remainder went to the C. Nash Herndon Fund, which still exists and is used for general operating expenses for the medical school.

Applegate said that the school would not take money from someone like Draper today.

"If there was one thing in this that surprised me, it was the leadership level involved in soliciting funds from Wickliffe Draper," he said. "That's sort of a difficult thing for me to accept but it happened. In hindsight, none of us would solicit money from that person."

The school now rigorously screens donors and their intentions before accepting money, Applegate said.

The committee also examined the role played by two doctors at the Bowman Gray School of Medicine. Dr. William Allan and Herndon taught and promoted the ideas of eugenics for years. Allan published many papers about the subject.

During his years as the chairman of the Department of Medical Genetics, Herndon wrote annual reports from 1943 through 1948 that referred to the department's eugenic activities, the committee found.

"Both Dr. Herndon and Dr. Allan supported the concept of eugenics, including involuntary sterilization and genetic counseling, and believed, as did many others during the early part of the 20th century, that it could be used to improve the health and welfare of society. Their status as faculty members provided a platform to advocate their views on eugenics," the report said.

Applegate said that similar programs could not exist today because of different attitudes about medicine and better safeguards against experimentation.

"In this time, as the report says, the ability to extrapolate from science without the checks and balances was greater," Applegate said. "In the modern paradigm the rights of the individual come first. Medicine is not to be used as a social tool."

Doctors at Wake Forest now have to go through the Institutional Review Board, an in-house committee composed of doctors, ethical and religious people and others, if they want to conduct any clinical trials or experimental medicine with human subjects, he said.

The report also found that faculty members performed sterilizations, including involuntary sterilizations, on patients. Committee members disputed that Herndon performed surgeries, saying that he was an internist and not a surgeon.

The Journal's series had said that Herndon performed at least six sterilizations, based on a comment attributed to Herndon in the minutes of the Human Betterment League recorded in November 1949. The minutes said that Herndon "had himself performed six operations in the past week and told of the very advanced policy of Baptist Hospital."

Committee members searched through the hospital records for six weeks before the statement and found no record of Herndon performing any surgeries, said Dr. Charles McCall, the chairman of the committee.

"We were convinced we could tell who the surgeon was. Dr. Herndon was not a surgeon and his name was not part of that (those records)," McCall said.

In the series, the Journal also reported that the medical school had collaborated with Forsyth County on the "Forsyth County Eugenics Program," which sterilized dozens outside of the eugenics laws.

Committee members said that they found nothing to indicate that the school's doctors had performed sterilizations outside of the law, which required that county workers petition for sterilization and a doctor sign an affidavit saying that the procedure was necessary. "There was just no evidence to suggest that among the documents we reviewed," McCall said.

Documents obtained by the *Journal* from the medical-school archives showed that Herndon worked with Forsyth County on a program that did not seem to be overseen by the state.

"In September 1943, a project aimed at eugenic improvement of the population of Forsyth County was begun in co-operation with Dr. J. Roy Hege, Forsyth County Health Officer. This project consists of a gradual, but systematic effort to eliminate certain genetically unfit strains from the local population. About thirty operations for sterilization have been performed," Herndon wrote in his annual report for the Department of Medical Genetics for the 1943-44 school year.

Until 1963, all sterilization operations in North Carolina had to be approved by the state eugenics board. There are no indications that the state approved nearly that many sterilizations in Forsyth County that year, according to Journal research.

Victims of sterilization are still waiting for help from state

September 5, 2004
By John Railey

Last year about this time, North Carolina seemed well on its way to making amends for the state-ordered sterilization of more than 7,600 adults and youth from 1929 though 1974.

Gov. Mike Easley had just approved a committee's recommendations that those sterilized receive health-care and education benefits, that a memorial be set up to them and that information about the sterilization program be included in the state's history curriculum. These recommendations and others were noble steps with noble goals – making amends and ensuring that programs similar to sterilization never occur again – and they were the first of their kind in the country.

Yet none of the recommendations have been carried out.

This after victims of the program including Nial Ramirez bared their souls to committee members, recounting memories of being sterilized that have long haunted them. "To me, it was a waste of time, because nobody did anything. It was a whole bunch of talk," Ramirez told me last week from her home outside Atlanta.

Carmen Hooker Odom, the secretary of the state Department of Health and Human Services, understands the frustration of Ramirez and other victims. "In their shoes, I would never have been as gracious as they were," Hooker Odom said. "I can understand their frustration. I'm very committed to getting this done."

For aging and hurting victims, time is dragging. And, in some cases, running out. At least a third of the victims are probably dead; others are elderly and infirm. Several, like Ramirez, are frustrated by the delay at easing the pain they've long felt.

The story she tells through tears is all too typical. Back in 1965 in Washington County, a social worker pushed her into agreeing to be sterilized at the age of 18, saying that her mother and the rest of her family would be taken off welfare if she didn't submit. By then, social workers who pushed the sterilization program had focused on poor black women like Ramirez. As they'd done when the program also included large numbers of whites, social workers pushed women and girls into sterilization by claiming that they were promiscuous or "feebleminded" - often based on flawed intelligence testing.

North Carolina's program operated beneath the radar screen and was largely unknown to the general public. After the Journal did a series on it in 2002, the state seemed on the road to speedy action to correct this massive injustice.

State House Rep. Larry Womble, a Democrat from Winston-Salem, called for reparations for those sterilized and continues to push for that. Several of the victims say money is the only thing that would ever come close to compensating them for taking away their God-given right to bear children.

But it's highly unlikely that reparations would ever be approved by legislators in this cash-strapped state. Given that, sterilization victims had taken some solace in the recommendations from the committee formed by

Easley.

State officials said the delays have been caused by several factors. The biggest factor is this: After exploring ways of carrying out the recommendations and hitting some road blocks, Hooker Odom said Easley told her the best way to get all the recommendations carried out is to make them part of a bill. "We've been doing a lot of ground work in trying to get information and talk to people and pull it all together, but sometimes you need a little heavier leverage in order to get people to understand that this is important to do," she said.

That bill couldn't be proposed in the N.C. General Assembly's short session that ended in July, so it will have to wait until the assembly reconvenes in January.

When the legislation passes, the state plans to form a panel to hear from sterilization victims and deliver them compensation.

If the legislation doesn't pass, I hope the state has a good backup plan for getting the sterilization victims health-care and education benefits and all the rest. Easley's initiative to help these victims was bold and courageous. The state was finally hearing the cries of folks to which it once turned a deaf ear.

The state must now complete this important work.

State drags feet on promise to sterilization victims

July 25, 2010
By John Railey

When the state of North Carolina moved to sterilize Willis Lynch in 1948 when he was 14, the process – from typing out the paperwork through the vasectomy – took less than a year. Most of the more than 7,600 victims of the state's forced sterilization program were sterilized quickly. But as the state considers compensation for the victims of the program that lasted from 1929 through 1974, it's moving on a much slower time clock. The state has been "studying" the issue for seven years, an agonizing delay that adds insult to injury for the victims of the program, many of whom were left with physical and emotional problems by the operations. "I believe they're waiting on us to die," Lynch, a 77-year-old retired handyman told me last week from his Warren County home.

North Carolina had one of the most aggressive sterilization programs in the country, ramping up even as other states backed away from the junk science of eugenics and ceased sterilizing their most vulnerable citizens. A Journal team of which I was part published a series in December 2002, Against Their Will, that revealed in unprecedented detail the workings of the program, which was aimed at "bettering" the state's human stock. The targets of the Eugenics Board of North Carolina were almost invariably poor, often sterilized for no more reason than they had had premarital sex or were labeled "feeble-minded" on the basis of faulty intelligence testing. Many like Lynch were residents of state training schools. At his school, Caswell Training School in Kinston, the Eugenics Board had decreed in 1935 that none of the inmates "should be released before being sterilized, except in the few instances where normal children had been committed by error."

After our 2002 series, Gov. Mike Easley promptly apologized for the program on the state's behalf and set up a blue-ribbon committee to study compensation. By the following August, he'd approved that committee's recommendations, the most important of which were health care and education benefits for the victims. But in five more years in office, Easley never pushed those benefits through the legislature. His successor, Bev Perdue, has yet to carry out her 2008 campaign promise to help the victims. She and Rep. Larry Womble of Forsyth County did get $250,000 included in the 2009-2010 budget to set up a foundation to study compensation for the victims.

That foundation, the N.C. Justice for Sterilization Victims Foundation, was slow to get started. In March, Charmaine Fuller Cooper was hired as the foundation's executive director at an annual salary of $55,465. But since then, Perdue has yet to implement her administration's plan to appoint a task force to work with the foundation. Cooper said the task force will make recommendations to the governor about "how to give any types of remedies to the victims, including compensation."

Cooper's office lacks its own website, which will be crucial in giving and receiving information to victims, their children, grandchildren and other relatives. All the foundation has is one page on the state Department of Administration's website, which it comes under. Cooper said she understands the frustrations of the victims, several of whom she has talked to. But setting up the foundation takes time, she said. She's had to work out many details, she said, such as how to ensure confidentiality for the victims.

Womble, long a champion for the victims, wants the process to move faster. Toward that end, he recently called a meeting with Cooper and Perdue. Womble said he made some requests to the governor and she honored them, including that a database will be set up to keep track of the victims and that allocations to the foundation will become part of her proposed budgets. The governor "looks forward to seeing progress soon" in helping the victims, a spokeswoman for her said.

But there's no clear plan for doing that. Cooper earned high marks as an advocate in her previous job as the head of the Carolina Justice Policy Center in Durham, but she needs Perdue's firm backing to succeed. North Carolina, which had one of the nation's worst sterilization programs, could become the first state in the nation to compensate its victims. Perdue should just embrace the enactment of the recommendations made by Easley's committee back in 2003, including health-care and education benefits for the victims, which could easily be done through the state's university system. She should back Womble's push in the legislature for those benefits, as well as financial compensation for the victims who are still alive. Get the compensation approved and let the foundation quickly dispense it.

The victims are dying off. And the living, forever changed by a quick operation, can't fathom why the state can't move nearly as quickly in helping them as it did in hurting them.

A state with a $25 million fishing pier can afford to help these victims

By John Railey
Journal Editorial Page Editor
May 29, 2011

Last weekend, Gov. Bev Perdue joined a crowd at Nags Head for the dedication of a state-operated fishing pier that was rebuilt for $25 million, mostly in state money, after being destroyed by Hurricane Isabel in 2003. Considering that amount of money could have compensated the living victims of North Carolina's forced sterilization program, you have to wonder what is wrong with this picture.

When the legislature and Perdue signed off on the pier in 2009, several of the victims had already long been making their case for compensation. Perdue had promised to help them as she ran for office in 2008. Other Democratic politicians have promised to help as well. The victims of this program that ran from 1929 through 1974, some of whom are hurting both mentally and physically, have heard eight years of hollow promises. Some have died waiting for help.

Some of the victims I often talk to say the legislature has to help them before its session ends this summer. "Maybe something will happen, I hope," said 77-year-old Willis Lynch of Warren County, a retired handyman who was sterilized in 1948 when he was 14.

There is reason for hope. A few days ago, Republican Rep. Dale Folwell of Winston-Salem, the speaker pro tem of the state House, told me that he would support compensation for the victims. "I think what happened to these people is awful and wicked and should never be be repeated," he said.

And a Perdue spokeswoman indicated that the governor is open to considering help for the victims this legislative session. Several victims had hoped the legislature's Republican majority, which won power last fall, would succeed where the Democrats have failed. John Hood, the head of the fiscally and socially conservative John Locke Foundation, has long pushed for compensation. "I think that state lawmakers see 2011 as the wrong year to authorize compensation because of the budget gap," he said in an email for this column. "They are mistaken. This is a past-due bill for a class of victims who shouldn't have to wait another year."

This issue, compensating people who were robbed of their reproductive rights by the state, should be a bipartisan one. The program rendered barren more than 7,600 vulnerable victims of modest means, often bullying them into operations with trumped-up reasons as it sought to "better society" and thin the welfare rolls. The program was supported by prominent doctors and families in Winston-Salem and other cities across the state. The state should compensate the victims financially and offer them health-care benefits through the state's university system, which could be done relatively easily and inexpensively.

Rep. Paul Stam, the House majority leader, had suggested that $20,000 for each living victim could be taken from money the state gets as its share of the national tobacco settlement. But the House budget rolled out recently doesn't include that idea, although it does include using the settlement money for other purposes.

Folwell said he doesn't know the mechanics of how compensation would take place. He thought Stam was figuring that out, he said. Stam told me several weeks ago that if his plan for using tobacco-settlement money to help the victims didn't work, he had a couple of other ideas. He has declined to disclose them. In response to repeated requests for comment over the last several days, one of his assistants finally sent me a one-sentence e-mail: "Rep. Stam asked me to let you know that there is nothing new to report as of now."

Try conveying that to the victims. "It's the same old thing, over and over," Lynch said,

The process of sterilizing them went quickly – usually no more than a year from the time a bureaucrat filed the first paperwork until they were wheeled into an operating room, often not told that the procedure would render them unable to reproduce. Some were told they were having appendectomies. The state is taking a lot longer to help them than it did to hurt them.

Perdue finally has a committee studying compensation, but its completion deadline falls long after when the legislative session should end. And the matter has been studied at length by two previous committees. In their records is all the evidence needed to justify compensation.

A state official told the latest committee last week that identifying people owed compensation could be a problem because many of the files don't have key biographical information. The program's lack of recordkeeping in that regard compounds the wrong. But Lynch and several other victims have shared their state sterilization records with the Journal, and those and any other victims whose records can be found and verified should be quickly compensated.

Chrissy Pearson, a spokeswoman for the governor, said that Perdue feels passionately about the issue and "is particularly concerned that these victims are aging and deserve justice sooner rather than later. It will remain on her list of items to consider with the General Assembly as they work through the upcoming budget."

Perdue and her fellow Democrats should make compensation a bargaining chip in the budget process. She wants the tobacco-settlement money to continue to go to economic development. But if she got behind the Republicans using a limited portion of it for other uses, they might just get behind her on using another portion of that money for sterilization compensation. And she could work with them to secure the health-care benefits.

Even in tight economic times, a state that can put millions into rebuilding a fishing pier can find a way to compensate the aging and dying victims of a program it should never have started. "They're dragging their feet," Lynch said.

Compensation: Help for sterilization victims past due

Journal Editorial, January 6, 2012

Sixty years ago on this page, we routinely supported our state's forced sterilization program. In 2002, as the Journal revealed the inner workings of that program in the investigative series "Against Their Will," the paper apologized for that support. Then-Gov. Mike Easley apologized on the state's behalf.

These apologies alone are not enough. For the state, the apology should come with compensation for the living victims of the program, which sterilized more than 7,600 men, women and children from 1929 through 1974 as part of its wrongheaded efforts to "better society" and thin the welfare rolls. As many as 3,000 victims may still be alive. For the past several years, we've backed our apology with editorials and columns pushing the state to finally help the victims instead of just making hollow promises to do so.

Now that compensation is closer than ever, as it should be, because many victims are still suffering from mental and physical ills left by their operations. Other victims are dying off. Reporters from around the world are watching, as North Carolina just could become the first American state to compensate sterilization victims.

With the legislature set to convene in May, the leaders of the state's GOP-dominated legislature, Thom Tillis and Phil Berger, have each gone on record with us as supporting compensation.

Democratic Gov. Bev Perdue's task force on compensation will soon release its recommendation on what that figure should be – probably at least $20,000.

Victims tell us $20,000 would be a slap in the face for being robbed by the state of their God-given rights to reproduce. But critics say even $20,000 – which, if there are as many as 3,000 living victims, would bring the tab to $60 million – is too much.

But we contend that a state that can find most of the $25 million for the rebuilding of a hurricane-damaged fishing pier at Nags Head – a deal Perdue signed off on – can sure find money to compensate these victims. And at least $20,000 for each victim should be paired with health-care benefits for the mental and physical ills left by the operations. The new year has started, and the legislature will be back for its regular session before we know it. Now is the time for the Republican leadership and Perdue to sit down together and map out a framework for finally doing right by these victims who were so terribly wronged.

Written by John Railey, Editorial Page Editor

Compensation: Use momentum to map out plan
Journal Editorial, January 22, 2012

State House Speaker Thom Tillis on Wednesday ramped up his commitment to compensating the victims of the state's forced sterilization program. "If we fail to ratify legislation in the short session," he told the *Journal's* editorial board, "I will consider it a personal failure." Momentum is building toward compensation.

We often disagree with Tillis on this page. But he is right in his commitment to this issue. The state needs such commitment from every member of the legislature to finally make compensation a reality. The victims are left with mental and physical ills from their operations. Some have died waiting for help.

Gov. Bev Perdue's task force on the issue has recommended $50,000 in compensation for each living victim of the program, which ran from 1929 to 1974. Tilllis' counterpart in the Senate, President Pro Tem Phil Berger, is on record as supporting some form of compensation. The legislature convenes in May, and Tillis said he wants a bill for compensation filed the first week.

That's good. But much groundwork needs to be done before then. Tillis said he has already started on that. The Republican said he wants to work with the Democratic governor and with Democratic Rep. Earline Parmon of Winston-Salem, among others, on the compensation effort. Parmon stands ready to ramp up the effort. Tillis' No. 2 man, House Speaker Pro Tem Dale Folwell of Winston-Salem, said Tuesday that he will help map out a compensation plan.

But while both the GOP leadership and Perdue support compensation, no one really knows where all the rank-and-file legislators of both parties stand. The issue has never made it to the legislative floor, despite years of trying by Democratic Rep. Larry Womble of Winston-Salem. He became the legislature's champion for the victims soon after the Journal revealed the sterilization program's inner workings in the 2002 series "Against Their Will." With Womble facing a long recovery from a December auto accident, new leaders are needed in the fight. Bipartisan cooperation is as rare in Raleigh as it is in Washington. But if ever there was a cause that our state leaders should unite on, this is it. This is about righting the wrongs of a government program that pushed itself into the private lives of more than 7,600 victims and their families, violating the sanctity of their lives and robbing us all of the promise of their progeny.

This program ended just a generation ago. It's still very much with the victims, many of whom are haunted by memories of the operations and plagued by physical ills left by them. We can't, as a state, turn our back on them any longer. Our leaders must come together on this one.

Written by John Railey, Editorial Page Editor

Tillis, Perdue Could Make History Together
By John Railey
Journal Editorial Page Editor
January 29, 2012

State House Speaker Thom Tillis, under attack by many liberals for leading the way in severe cuts to public education, is on the verge of making state history in a positive, bipartisan way -- with the Democratic governor who has lashed out at him on those cuts.

If his plan works, this coming session of the legislature will finally approve compensation for the living victims of the state's forced-sterilization program that hustled more than 7,600 men, women and children into operating rooms from 1929 through 1974 to "better society" and thin the welfare rolls. The state that ran one of the most aggressive sterilization programs in America would make worldwide headlines as the first state to compensate sterilization victims.

That is, if the Republican speaker can overcome several big hurdles, the largest of which is cooperating with Gov. Bev Perdue, who has been working on her own plan for compensation. That part may have gotten easier with Perdue's announcement last week that she is not seeking re-election. Before, Republicans might have been scared of helping her achieve compensation, fearing she might claim it as a political win. She might have felt the same way about Republicans. Now, hopefully, both sides will work together for the victims.

Victims such as 73-year-old Annie Buelin of Surry County hope the politicians will finally come through for them. "I just think they need to have some common sense and try to put themselves in my place, to see how they would feel if it happened to them," Annie, who was sterilized when she was 13, told me last week.

The other day, I talked for about an hour with Tillis about his plan. "If we fail to ratify legislation in the short session," he said, "I will consider it a personal failure."

And Tillis does not like failure. A successful businessman, he rose to the top of the House in just a few terms. But some in the old guard of his own party resent his rapid rise and aggressive style. He realizes that he needs their support, as well as Democratic support, to get compensation passed.

"One thing I'm really trying to do is not make this a political thing," Tillis said. "I want to manage the emotions around it. ... I want to manage this very tightly. So we don't get distracted and fail to deliver." Despite years of trying by Rep. Larry Womble of Winston-Salem, who championed compensation soon after the Journal in 2002 revealed the inner workings of the sterilization program, the issue has never made it to the House floor. And if Tillis succeeds in getting the House to approve it, the Senate still has to approve as well.

Phil Berger, the Senate president pro tem, told me several weeks ago that he supports compensation. Tillis has been far more vocal about compensation. He and Berger are not particularly close. And if the House and Senate approve compensation, Perdue would still have to sign it into law, provided both chambers capitalize on their momentum and get it to Perdue before she leaves office in December.

Tillis said he plans to talk to Perdue "with an eye toward having a consensus" on sterilization compensation.

In its final report released Friday, Perdue's task force repeated its recommendation of $50,000 for each living victim. Tillis won't yet say what he thinks the figure should be. But he emphasizes that the victims definitely need to be compensated. Tillis is Catholic. But he said that "aside from the moral issues that this program raised – and they are numerous – I've tried to really emphasize the argument of government taking from somebody the most precious piece of property they have, their body, or parts of it. I've tried to articulate this as an example of an egregious violation of people's personal property rights. If you could rationalize a program like this... that ultimately takes away a part of somebody's body, where does it end? You could rationalize that government should be able to take anything away."

Tillis has met and listened to victims such as Annie Buelin. "I just admired her grace," Tillis said of Annie. "I just thought that she and most of the victims that I have met or heard stories of, it's amazing to me the grace that's exhibited by the victims themselves."

Tillis said he wants to finally help those victims. "The main thing is to bring some finality to this process."

Sterilization Compensation: It fits with state's character
Journal Editorial, February 3, 2012

By and large, most North Carolinians are against big-government intrusion into their private lives. They also believe in the sanctity of human life, and protecting our most vulnerable citizens from government abuse. That is one big reason why it's crucial that the legislature, when it convenes in May, quickly pass $50,000 in compensation for each living victim of this state's forced sterilization program.

And it's crucial that Gov. Bev Perdue give her input so that the legislature will produce a compensation bill she can ratify. Victims such as Bertha Dale Midgett Hymes, who was sterilized in 1967 in Dare County when she was 17, say they've waited long enough.

"They need to help us," Hymes said this week.

This state, which ran one of the most aggressive sterilization programs in the country, should become the first state to compensate sterilization victims.

It remains a mystery why supposedly progressive North Carolina ever bought into the nationwide sterilization movement, aimed at "bettering society," in the first place. But buy in it did, giving its social workers unique powers to target for sterilization groups including the blind, the epileptic and the "feeble-minded" – a catch-all designation often based on flawed IQ testing. It zeroed in on black women and girls in the 1960s, when most other states had backed off their sterilization programs. But make no mistake: The victims also included large numbers of whites, as well as some American Indians. Almost all of the nearly 8,000 victims rendered barren from 1929 through 1974 shared one common denominator: They were poor and powerless to fight back against a program supported by prominent families and doctors in Winston-Salem and across the state.

The program, called the North Carolina Eugenics Board, came under the state Department of Public Welfare. The sterilization program should have answered to our governors and our legislators. But they failed to monitor it, and the program became a dangerous power unto itself. It hid in plain sight, even claiming to be a positive force because, it said, it stopped "morons" and other "undesirables" from reproducing. The program's brutal inner workings were not revealed until the *Journal* published the investigative series "Against Their Will" in late 2002. Ever since, politicians have promised help. Victims, quite understandably, are skeptical now that help will ever come.

The delay is shameful. Some victims have died waiting for help. Others continue to suffer mental and physical ills left by the operations. After telling their stories to us, they've told them to reporters from across the nation and overseas.

They've rightly borne witness to a horrible chapter in our state history. We can't close that chapter until we compensate the victims, showing the world that we North Carolinians own up to our mistakes and learn from them.

Help for sterilization victims such as Bertha Dale Hymes is overdue.

Written by John Railey, Editorial Page Editor

Credits

In the summer of 2002 the *Winston-Salem Journal* started examining the Eugenics Board of North Carolina and the state's sterilization program after gaining access to a cache of non-public documents. Several members of the Journal's staff were pulled from their regular duties to focus on the issue. The reporters traveled to several states and across North Carolina to conduct interviews and do research.

Kevin Begos was the *Journal's* Washington, D.C. reporter. He has also reported from Sudan, Afghanistan and Iraq. Begos developed the lead and created the foundation for the series, and has continued to write about the eugenics movement. He is a former Knight Science Journalism Fellow at MIT and a graduate of Bard College. He has won national awards from the Associated Press Managing Editors and *Washington Monthly*.

Danielle Deaver covered medical affairs for the *Journal*. A 1998 graduate of Wake Forest University, Deaver started at the *Journal* in 1999. She has written about stem-cell transplants and a hyperbaric oxygen chamber used to treat brain-damaged children.

John Railey has won numerous awards for his writing, including the national Cornell Memorial Award in 1999 for religion coverage. Railey became an editorial writer at the *Journal* in 2004 and the editorial page editor there in 2010. In both capacities, he has made the *Journal* the leading voice for compensation for the sterilization victims. He is a 1984 graduate of the University of North Carolina at Chapel Hill.

Scott Sexton was the project's editor. He then continued to speak out about the sterilization program as a columnist for the *Journal*. He joined the paper in 1999 as night editor and is a graduate of the University of North Carolina at Greensboro.

Ted Richardson was a photojournalist at the *Journal*. He is a graduate of Davidson College and has a master's degree from the University of North Carolina at Chapel Hill. Richardson has won several awards for his photography.

• Katherine Elkins and Jennifer Falor, interactive media producers for Journalnow.com, designed and produced the online series and its components.

• The series was copy-edited by Jo Dawson. Charlie Elkins, an assistant managing editor, designed and paginated the pages for print. Jana Dwyer provided research help. The graphics were done by Jim Stanley and Nicholas Weir. Michelle Johnson produced the online audio collected by Ted Richardson.

The *Against Their Will* project won awards from Investigative Reporters and Editors, the Newspaper Guild of America, the Society of Professional Journalists, the Online News Association, the National Association of Black Journalists, the American Legion Auxiliary, and the Association of Women in Communications.

INDEX

A

Abortion: and eugenics, 83, 117, 128, 129, 138, 163, 172,

Academia: role of, 135, 139

Advertising and eugenics: 71, 77

Allan, William: 36, 37, 38, 39, 41, 42, 44, 45, 61, 62, 63, 146, 217

African Americans: eugenics and, vi, 1, 2, 3, 8, 16, 30, 49, 50, 52, 53, 55, 57, 58, 69, 70, 73, 76, 77, 79, 81, 88, 89, 91, 103, 104, 105, 106, 107, 108, 116, 117, 123, 124, 125, 133, 136, 137, 140, 152, 157, 161, 164, 172, 185, 189, 194, 206, 216, 220

Birth control: 4, 10, 26, 34, 72, 73, 74, 75, 77, 79, 118, 141, 142, 143, 145, 146, 147, 163, 164, 165, 173

Aid to Dependent Children (ADC): 48, 105

B

Begos, Kevin: v, 1, 15, 29, 33, 34, 47, 55, 69, 71, 93, 109, 119, 141, 157, 160, 166, 170, 181, 187, 232

Birth Defects: 45, 132, 146,

Blacks (see African Americans)

C

Castration: 69, 70, 174

California: eugenics and, 5, 136, 157, 161, 169, 178, 181, 182, 183, 199

Caswell Training School: 99, 100, 119

Cox, Nial: 2, 47-53, 158, 161, 175, 184-188, 190, 192, 195-197, 201-203, 208, 209, 219, 220

D

Deaver, Danielle: v, 25, 35, 61, 87, 127, 135, 145, 149, 157, 160, 163, 170, 207, 210, 215, 232

Draper, Wickliffe: 55-59, 216, 217

E

Easley, Mike: 157-161, 166-171, 179-181, 184, 186, 188, 189, 191-193, 195-197, 199-201, 207-209, 219, 221

Epilepsy, eugenics and: 3, 9, 69, 111, 178

Eugenics, apologies for: i, 24, 157-159, 162, 169, 181, 183, 210

F

Feeblemindedness: 16, 19, 24, 30, 31, 42, 61, 66, 73, 77, 95, 96, 106, 100, 111, 119, 166,

Forsyth County: 2, 35-37, 44, 71, 79, 106, 155, 218, 219

G

Gamble, Clarence: 3, 72, 73, 74, 77, 78, 88, 89, 97, 142, 145, 146

Genetics, role of: 35-45, 55, 56, 61, 72, 78, 79, 90, 116, 130, 136-138, 146, 216, 217, 219

H

Hanes, James: 3, 71-79, 87-89, 153

Herndon, C. Nash: 35-44, 56, 59, 62, 63, 72, 77, 78, 88, 116, 216-219

Human Betterment League of North Carolina: 3, 37, 39, 71-79, 95, 96, 120, 142, 153, 154, 173, 215, 216, 218

Hymes, Bertha Dale: 81-86, 158, 161

I

IQ tests: role of, 18, 19, 23, 31, 62, 73, 77, 81, 83, 87-91, 93, 120, 136-138, 170, 173-175, 187, 189, 192, 205, 206, 210

J

Jessie, Elaine Riddick: 15-25, 158, 170, 174-177, 184-190, 192, 195, 196, 197, 202, 203

Jolly, Wilbur: 103, 104, 107

K

Koomen, Jacob: 19, 22, 23, 30

L

Lawrence, George H.: 75, 75, 163

Lawsuits, sterilizations: 21, 30, 52, 83, 96, 161, 162, 175

Lombardo, Paul: i, 3, 5, 6, 33, 56-58, 162, 167-169, 182, 183

M

Mental illness: and eugenics: 2, 11, 69, 71, 111, 128, 172, 175, 178, 180, 187, 189, 192

Moore, Ernestine: 203-207

N

Native Americans: and eugenics, 116, 123-125

Nazi Germany, eugenics and: 2, 24, 35, 55, 57, 103, 132

Newspapers: role of, 3, 13, 30, 93-96, 153

North Carolina State Archives: 4, 33, 69, 159, 167, 168, 172, 201, 202, 205

North Carolina, University of: 33, 69, 72, 87, 94, 105, 120, 167

Nurses: role of, 51, 73, 106

O

Oregon: eugenics and, 5, 6, 156, 169

Operations: deaths from, 166-169

Operations: use of radiation in, 166, 167

P

Poverty: and eugenics, 16, 51, 79, 85, 94, 164, 206

Peru: sterilizations in, 34

Pioneer Fund: and eugenics, 55-58, 136

Poteat, William Louis: 65-67

Perdue, Beverley: 220-230

Q

Quinacrine: 141-143

R

Racism: and eugenics, 55, 58, 104, 139, 140, 194

Railey, John: 15, 29, 47, 65, 81, 103, 109, 113, 115, 123, 157, 160, 170, 184, 191, 203, 207, 219, 222, 225, 228, 230

Ramirez, Nail Cox: 2, 47-53, 158, 161, 175, 184-188, 190, 192, 195-197, 201-203, 208, 209, 220

Richardson, Ted: 232

Riddick, Elaine (see Jessie, Elaine Riddick)

Riddick, Tony: 24, 103, 186, 196, 197

Rushton, Philippe: 58, 59, 137, 139, 140

S

Schoen, Johanna: 4, 10, 11, 23, 33, 36, 37, 50, 74, 77-79, 142, 143, 167, 168, 194-196

Sexton, Scott: 232

Social Workers: role of, 12, 30, 49, 50, 53, 75, 81, 82, 105, 106, 110, 114, 119, 121, 126, 172-174, 185, 210, 220

Stonewall Jackson Training School: 109-112

T

Tubal ligation: history, 25, 26

W

Wake Forest University: role of, 1, 2, 35-37, 55-57, 61, 65-67, 117, 157, 195, 215-219

Winston, Ellen: 75, 120, 172

Womble, Larry: 162, 178-180, 186-190, 192, 193, 196, 198, 199, 202, 203, 209

APALACHICOLA MUNICIPAL LIBRARY
74 6TH STREET
APALACHICOLA, FL 32320

Against their will: North Carolina'
613.9 BEG   39026002162634

Apalachicola Municipal Library